COMMUNICATIVE PRACTICES
IN WORKPLACES AND
THE PROFESSIONS

Cultural Perspectives on the Regulation of Discourse and Organizations

Edited by
Mark Zachry
University of Washington

Charlotte Thralls
Western Michigan University

Baywood's Technical Communications Series
Series Editor: CHARLES H. SIDES

Routledge
Taylor & Francis Group

LONDON AND NEW YORK

First published 2007 by Baywood Publishing Company, Inc.

Published 2017 by Routledge
2 Park Square, Milton Park, Abingdon, Oxfordshire OX14 4RN
711 Third Avenue, New York, NY 10017, USA

First issued in paperback 2017

Routledge is an imprint of the Taylor & Francis Group, an informa business

Library of Congress Catalog Number: 2006032749

Library of Congress Cataloging-in-Publication Data

Communicative practices in workplaces and the professions : cultural perspectives on the regulation of discourse and organizations / edited by Mark Zachry and Charlotte Thralls.
    p. cm. -- (Baywood's technical communications series)
  Includes bibliographical references and index.
  ISBN-13: 978-0-89503-372-7 (cloth : alk. paper)
  ISBN-10: 0-89503-372-0 (cloth : alk. paper) 1. Business communication. 2. Discourse analysis. I. Zachry, Mark, 1968- II. Thralls, Charlotte. III. Series.

  HF5718.C6424 2007
  302.3'5--dc22

                      2006032749

ISBN 13: 978-1-138-63741-2 (pbk)
ISBN 13: 978-0-89503-372-7 (hbk)

Cover photograph by Christopher League

Table of Contents

PART I:
Understanding Regulative Processes,
Practices, and Effects

PART II:
Regulation and the Possibilities of Action:
Agency, Empowerment, and Power

PART III:
Critical Research Perspectives

Regulation and Communicative Practices

Mark Zachry

In recent years, scholars have become increasingly interested in studying routine or regularized discourse and its connections to the many institutions within which people act. This interest has emerged for a number of reasons, including the rapid evolution of digitally mediated communication and the upheaval of organizational structures in such realms of activity as business, government, and education. At the same time, theories of communication have become increasingly complex, beginning to account for such considerations as power, contingency, and unpredictability and their relationship to the practice of routine or regularized communication. As we observed research interests developing along these lines and attempted to understand the ideas at play in the related scholarship, it became increasingly evident to us that the moment had arrived for organizing a collection that begins to account for routine or regularized communicative practices in workplaces and professions.

Scholars publishing on this issue work in diverse fields of study (e.g., rhetoric, cultural studies, and organizational communication), and often use distinctly different theoretical perspectives (e.g., genre, narrative, and ethnomethodology). Consequently, what we perceived as an important, emerging concern in communication scholarship (in the broadest sense) lacked a named, central concept that would readily cross fields of study and allow us to talk about the enterprise at large. Recognizing that any starting point for organizing talk about this as-yet unnamed concern in communication studies would necessarily be somewhat arbitrary, we decided to bring together scholars situated in varied disciplines and theoretical camps and to ask them to grapple with what we proposed to be a central concept that seemed to unite aspects of their scholarly work. The concept we proposed was regulation. It was a concept that was not immediately identified with the work of any of these scholars, which had the benefit of placing all the contributors in roughly the same situation: aligning their thinking around a concept that had potential but unarticulated connections to their earlier work.

REGULATION AND COMMUNICATIVE PRACTICES

For many readers, the verb regulate is probably most immediately associated with rulemaking by a governmental or administrative body. A fundamental pair of activities for such bodies, after all, is to create and enforce regulations. It is commonplace, then, to associate regulations with either the law or administrative policies as the authorized, explicit controls imposed by an entity that is constituted for such purposes. In this context, regulations have a clear and often direct connection with the communicative practices of workplaces and professions. For example, any given business is required to report certain types of information to appropriate regulating agencies, or a credentialing board may require that the individuals it certifies document their yearly activities.

While these obvious and widespread forms of regulation are important and merit scholarly attention, the activity of regulation as it has emerged in recent scholarship is focused in other, less apparent areas, too. Beyond the explicit controls of governmental and administrative bodies, a complex configuration of factors exists that orders the communicative practices in which people in workplaces and professions engage. These factors include such things as groups of people regulating their own speech and writing to conform to professional norms, as well as individuals tacitly agreeing to suppress their opinions to avoid conflicts that could jeopardize their jobs. Examining how communicative practices come to be self-regulated by factors like these is as important as understanding those regulatory controls explicitly imposed by an external body because both always work concurrently. That is, communicative activities are regulated in multiple, interconnected ways. This multiplicity and how it works, however, has emerged as a fairly recent concern for communication scholars—a development that we briefly overview later in this introduction.

Yet another fundamental consideration in the study of regulation and communicative practices is that the relationship between the two is not unidirectional. Regulation is not something that is imposed on communicative practices by an external force, regardless of whether that force is an authorized body such as the government or the communicator herself. Regulation is always itself constituted and sustained by communicative practices. That is, the existence of regulation is contingent upon the communicative practices that define and enable it, just as the existence of those practices is contingent upon the regulatory forces that make them meaningful. The ratios and workings of this relationship, however, are only scarcely discussed by scholars—a situation that this collection is designed to help address.

Together, the contributors to this collection offer a provocatively complex picture of what regulation means and the means of regulation. The workplace or professional sites that the authors use as illustrations in their studies are diverse, covering such organizations as an Internet start-up company, an international energy company, an urban hospital, a university, and a telecommunications corporation. The perspectives the contributors bring to bear on their work cover a range of prominent thinkers, including sociological theorists (e.g., Bourdieu, Giddens, & Latour), philosophers (e.g., Habermas, Foucault, & Deleuze), and textual theorists (Spivak, Bakhtin, & Burke). In total, the perspectives offered by these contributors are invaluable for researchers who want to gain greater insight into routine or regularized discourse and its connections to the many institutions within which people act.

THE EMERGENCE OF SCHOLARLY ATTENTION BEING PAID TO REGULATED COMMUNICATIVE PRACTICES

Over the last 25 years, a large and diverse group of communication scholars has been exploring relationships between workplace and professional communication and an ever-expanding constellation of factors that help account for how that communication is practiced. Working from different theoretical and methodological perspectives, scholars engaged in this exploration have largely worked without the benefit of a basic set of shared terms that would bring some unity to their many invaluable but disparate projects. As I indicated earlier, a primary purpose for this collection is to begin putting some shape to this exploration.

The idea of regulation, when it is defined along the lines of relationality and contingency described above, has a somewhat shorter lineage in communication studies. It emerges in the work of people thinking about communication from a social constructionist perspective in the late 1980s. In these early instances, some overarching framework is proposed as central to understanding how communicative practices are regulated (or, to use a common term in this work, shaped). So, for example, a given profession might be advanced as the central explanatory feature of a certain type of discourse (e.g., Selzer, 1983). Likewise, a given organization could serve a similar role (e.g., Doheny-Farina, 1986). Organizational context, in fact, became a dominant explanatory feature for understanding the practice of professional communication (e.g., Driskell, 1989). At the same time, scholars were pushing for increasingly complex understandings of context (Harrison, 1987).

Some of those whose work was associated with social constructionist perspectives, for example, were increasingly likely to expand the idea of context so that its implied boundaries became more and more difficult to imagine. Pare (1993), for one, interrogates the idea of a discourse community, observing that many factors account for how discourse is regulated. Only some of these factors, he demonstrates, are accounted for in "standardized documentation," while other factors are implicit (p. 112). As he explains, "the dynamics of discourse regulations" can be understood through a "web of relationships," which extends through "complex connections" (pp. 114-115). In similar fashion, Porter (1993) argues that "the significant parameters" that need to be accounted for to understand professional writing "will be neither the discipline nor the corporation—but rather . . . a diverse network of concerns, extending among corporations, disciplines, and citizens" (p. 141). This network of concerns includes such extra-organizational considerations as cultural values, ethical/moral codes, and public policy and law (p. 142).

The few pieces cited here are suggestive of a larger trend in communication studies to turn toward increasingly complex accounts of why communicative practices exist as they do. These accounts were called by many "social" (see, for example, the overview of "the social perspective" in Blyler & Thralls, 1993), as communication scholars worked in and around the current of social-constructionist thinking. Such thinking is still being productively pursued (including by some participants in this collection), though many have over the last decade opted to abandon the social as an organizing principle because they want to move beyond the essentialist associations that many people now assign to social constructionism. Rather than arguing for a refined definition of the social, these scholars turn toward nonessentialized conceptions of culture to account for how language and other practices function in human experience. For many, this turn toward a cultural perspective allows researchers greater opportunity to consider the contingent and always unpredictable play of such factors as gender, ideology, and history in how communication is practiced. This turn has also brought with it (and enabled) new ways of conceptualizing communicative practices, many of which are represented in this collection.

KEY IDEAS IN THE STUDY OF REGULATED COMMUNICATIVE PRACTICES

In their efforts to make sense of how communicative practices are regulated in the sense suggested above, scholars have drawn new attention to issues that were often elided or oversimplified in earlier communication studies. In particular, these scholars have all pointed to inescapable problems associated with treating communicative practices as discrete, codifiable events in which a known set of elements (e.g., text, author, and audience) operates. As these new scholars insist, any boundary drawn around a communication event is always artificial, inevitably ruling out some thing(s) that interacts with whatever has been ruled in. For this reason alone, it is always therefore impossible to isolate all the elements that are at play in the practice of communication. Or, for that matter, it is likewise impossible to assign an autonomous and definable existence to elements that have been identified.

The quandary for scholars working from this position, then, is how to talk at all about communicative practices as they exist in the world. If, as these scholars assume, the discrete elements involved do not offer sufficient explanatory power for understanding how the practices of communication are regulated, then what are we to make of these elements? And, more importantly, where might we next turn to understand regulation? Rather than overly emphasizing the importance of constituent elements in the practice of communication, scholars who are concerned with regulation tend to foreground whole other sets of issues in which such elements are only of marginal significance. Focusing on whatever it is that makes the practice of communication regular rather than on the elements of artificially defined communication events, these scholars tend to be concerned with three broad, overlapping sets of issues: relationality, situatedness, and agency.

Relationality

A key consideration in studying the regulation of communicative practices is accounting for how those practices relate to the order of things around them. For example, a scholar studying communicative practices in an organization is likely to first look to in-house protocols, style guidelines, habits, etc. when attempting to understand how members of that organization communicate with one another and external entities such as government bodies or the public. Similarly, a scholar attempting to account for the communicative practices of a professional group might consult relevant professional journals, technical dictionaries of language and usage, and even academic textbooks to understand how members of that profession initially learn and then become enculturated to communicate in the ways that they do. A general movement in such scholarship, however, is the growing realization that a simple theory of relationality is inadequate for coming to a deep understanding of the regulation of communicative practices. That is, when a scholar attempts to map causal relationships between the communicative practices of a workplace organization or a profession and such things as a published style guideline or a technical dictionary, communication as it is practiced always exceeds or otherwise defies the explanatory power of the object it is being mapped against. Ultimately, then, scholars who are attempting to form a deep understanding of communicative practices tend to turn toward increasingly complex theories of relationality.

A central premise in these complex theories of relationality is contingency. In different ways, many communication scholars working in this area now operate from the premise that communicative practices are always *possibly* but also always *uncertainly* accounted for in their relationship to the order of things around them. In other words, communicative practices do not stand in a logical and necessary relationship with any external entity, though in any given instance such a relationship may be at play. This realization manifests itself in different ways in communication scholarship, ranging from strongly hedged claims in an account of how certain communicative practices evolved to studied examinations of contingency at play in a given organization's corpus of public texts.

Situatedness

For scholars attempting to understand how communicative practices are regulated, the issue of situatedness is another prominent consideration. As I mentioned earlier, much of the early work on this issue was framed in terms of studying context where the term *context* served as a sort of catchall for all those things that were known to surround a communication event but that could not be isolated as an integral element of the communication event. In other words, context stood in as a reduction for everything that could not be counted as the communication itself or as the originator and receiver of the communication. The challenge for scholars using this metaphor was twofold: first, the lines drawn between the communication event and everything else around it proved to be notoriously difficult to draw in a meaningful way, and second, it was impossible to reasonably bound what does or does not count as the context for a given communication event. Consequently, to invoke the idea of a context that exists separately from communication, but yet could provide an adequate explanatory framework for understanding how an act of communication was regulated, has proven for many scholars to be unsatisfactory.

Steering away from the problematic and ultimately misleading metaphor of context, many communication scholars now prefer more complex metaphors such as networks, open systems, and rhizomes for talking about the situatedness of communicative practices. Theorizing communicative practices is no less difficult when using these alternative metaphors, but these metaphors have the advantage of seeming less fixed than context. That is, these metaphors emphasize the recombinant, extensible, and expansive situatedness of communicative practices, allowing researchers to avoid the subject/object or center/periphery relations that are associated with studies focused on context. For this reason, these alternative metaphors have proven to be more engaging tools for thinking about the regulation of communicative practices, calling attention, in complex ways, to such interrelated ideas as time, space, material resources, and power.

Agency

A final, overarching issue of concern for scholars considering how communicative practices are regulated is agency, or the identity position occupied by the participants. From a humanist perspective, people involved in communication are traditionally assumed to be autonomous, self-regulating individuals. A skilled author, for example, is someone who is disciplined in the art of crafting texts. Likewise, a reader is treated as an individual capable of drawing upon her own ideas to critically receive and measure the words of an author. For various reasons and in many different ways, most scholars thinking about communication have abandoned such traditional humanist conceptions of agency. In particular, most communication scholars now operate with a more complex understanding of individuality, generally accepting the idea that people do not operate as autonomous, self-regulating entities. Instead, people are understood to be complex sites of conflicting social, biological, educational, and other materially conditioned factors that are not of their own devising. In addition, people are always positioned in ways that limit their apparent volition. For scholars, these limits challenge the extent to which people may be said to possess agency when they are engaged in communication practices. Obviously, then, questions about who and what is regulating communicative actions become more complicated.

For many scholars, this line of thinking has migrated toward a posthuman perspective on communication. No longer are individuals assumed to be the sole possessors of the agency and intelligence that regulates communicative practices. Instead, agency and intelligence are treated as properties distributed between humans, machines, and interfaces, thereby essentially displacing human actors to more marginal positions. At the same time, though, there are many other scholars who prefer to continue working in the humanist tradition by arguing for more sophisticated definitions of what a human is and how agency is constituted.

As readers of this collection will note, the issues of relationality, situatedness, and agency can be explored in many different ways, but they are never too far removed from central concerns of scholars in this area.

OVERVIEW OF THE CHAPTERS

The participants in this collection offer varied perspectives on the idea of regulation and its relationship to communicative practices in workplaces and the professions. For some, the idea of regulation serves as a central organizing concept for their chapters; for others, regulation functions as a secondary concern for illuminating a related theoretical perspective or illustrative research study. Chapters in Part I of the volume are primarily concerned with helping readers understand regulative processes, practices, and effects. In Part II, then, the focus shifts to considering agency and power in terms of their connection with regulation. Finally, in Part III, the chapters focus primarily on offering research perspectives that have the potential to influence how scholars critically study regulation and communicative practices.

Dorothy Winsor begins Part I with "Using Texts to Manage Continuity and Change in an Activity System," which examines the centrality of texts in the regulation of human activities. As she demonstrates, the regulatory effects of texts are complicated by a number of factors including continuous change in the systems within which humans act and the competing goals of those who must act with and around the texts. Winsor supplements cultural/historical activity theory with the work of Bourdieu and Foucault to examine how the regulatory effects of charter documents are negotiated in subsequent, subsidiary texts.

In Chapter 2, "Regularized Practices: Genres, Improvisation, and Identity Formation in Health-Care Professions," Catherine F. Schryer, Lorelei Lingard, and Marlee Spafford study how students attempting to join health-care professions engage with regulated resources of their intended profession and with regularized ways of knowing within that profession. The authors draw a distinction between regulation, which they associate with external controls, and the regularized actions of humans, meaning the actions into which agents opt to order themselves. Using the work of Bourdieu, Giddens, activity theorists, and genre theorists, the authors examine how participating in the situated practice of case presentations regularizes professional identities for health-care students.

Clay Spinuzzi's Chapter 3, "Who Killed Rex?: Tracing a Message through Three Kinds of Networks," explores two prominent variations of sociocultural theory, activity theory and actor/network theory, to consider how the material regulation of human activities is accounted for in each theory. Using primarily the works of Engeström and Latour, Spinuzzi considers the concept of self-regulation in three networks, each providing its own perspective into the work of a telecommunications company. As he demonstrates, the ideas that might be commonly associated with self-regulation—agency, cognition, and responsibility—are differently configured depending upon the network perspective from which they are viewed. Such reconfigurations, he contends, are ultimately productive for researchers.

Recasting regulation in terms of constraint and enablement, JoAnne Yates and Wanda Orlikowski's Chapter 4, "The PowerPoint Presentation and Its Corollaries: How Genres Shape Communicative Action in Organizations," examines the shifting practices of conventional communication in the interplay of new technology and human improvisation with it. They introduce the concept of corollary genres to account for inflections in conventional discursive practice as people explore how a new technology can enable or constrain their communicative actions. Giddens' structuration theory is coupled with the work of genre theorists as the authors engage in a historical study of business presentations and then examine the uses of PowerPoint in different organizational contexts.

Chapter 5, Martin Ruef's "Reason and Rationalization: Modes of Argumentation Among Health-Care Professionals," considers rationality and its relationship to the regulation of professional authority in health care. After outlining an expanded definition of rationality, Ruef reports the results of a systematic, empirical analysis of modes of argumentation composed for different professional audiences in health care. He considers how factions of a given profession contest the regulation of their professional communication.

Kenneth Gergen's Chapter 6, "Writing and Relationship in Academic Culture," challenges the primacy of traditional and, therefore, conventionally regulated forms in academic writing. He notes the undesirable implications of such forms, arguing that they lead to isolated subcultures within the academic world. As an alternative, Gergen proposes relational genres of expression that expose the writer to the reading audience in unconventional ways. Developing a social constructionist approach to communicating via new forms of representation, he advocates that the regulatory force of traditional forms, particularly those that are perpetuated in the natural sciences, should be replaced by a new paradigm.

Leading Part II of the collection, which focuses more explicitly on questions of agency, empowerment, and power, is Carl G. Herndl and Adela C. Licona's "Shifting Agency: Agency, *Kairos*, and the Possibilities of Social Action." Looking at how social action is made possible in varied and changing social spaces and practices, Herndl and Licona develop a theory of agency that rejects the idea that agency is something that individuals can possess, arguing instead that agency is something that is enacted in the midst of shifting sets of social and subjective relationships. They theorize the idea of "constrained agency," which designates the time- and space-bound potential for agency in the presence of the regulating forces of authority.

Dave Clark's Chapter 8, "Rhetoric of Empowerment: Genre, Activity, and the Distribution of Capital," interrogates the very idea of worker empowerment, showing how power is always inherently regulated within organizational networks in ways that belie narratives of empowerment. Replacing the traditional binary explanation of organizational power with an explanation based on "the broad array of discourses, technologies, professions, traditions, and capital that regulate," Clark argues for a cultural approach to understanding what power is and how it operates in organizations.

Chapter 9, Barbara Schneider's "Power as Interactional Accomplishment: An Ethnomethodological Perspective on the Regulation of Communicative Practice in Organizations," advances the argument that conceptualizations of power as something that is simply possessed or inherent in a hierarchical position are misguided. Power, instead, is something that is accomplished by individuals engaged in the communicative practices of everyday social interactions. She contends that power is integrally connected to how the participants in social interactions regulate their communicative activities. And while these participants may orient their activities around "rules of particular genres of discourse," those rules in themselves do not regulate communicative practices.

The final Part of the collection focuses on critical research perspectives, beginning with Brenton Faber's "Discourse and Regulation: Critical Text Analysis and Workplace Studies." In this chapter, Faber offers a research perspective for how forces of change and resistance are regulated within workplaces. Specifically, he argues for examining macro and micro discursive features in organizational texts to arrive at a critical account of the play of regulatory actions.

In conducting a critical text analysis of an organizational effort to implement a new information technology platform, Faber examines "the social aspects of regulation and the ways regulatory discourse can take place as simultaneously multiple, coordinated discursive actions."

David M. Boje offers narrative theorists a new research perspective with "The Antenarrative Turn in Narrative Studies" in Chapter 11. Observing that the coherent structure of narratives (with their beginnings, middles, and ends) inherently regulates how something is known, Boje asks what researchers are to make of all the fragmented texts that circulate within organizations but are not regulated by narrative structures. Antenarrative, he suggests, provides a useful way of understanding and accounting for such stories that circulate before and around proper organizational narratives. Using events at Enron to illustrate his theory, Boje examines the multiple and conflicting antenarratives that complicate the official narrative of what happened with this failed company.

This final part of the collection concludes with Robert P. Gephart, Jr.'s "Hearing Discourse." Gephart advocates a critical/interpretive approach to analyzing regulated communicative practices such as those that occur at governmental hearings. His approach, which combines narrative/rhetorical analysis, ethnomethodology, and Habermasian critical theory of speech acts, offers insight into how organizations are produced in and through communicative practices. Gephart uses his study of public hearings related to a fatal oil-pipeline fire to illustrate how this perspective enables researchers to examine formalized communicative practices with new insights. Through this example, too, Gephart advances the argument that regulatory effects are not unidirectional.

The 12 chapters offered in this collection provide thoughtful perspectives on how we can understand the nature and work of regulation in the communicative practices of workplaces and the professions. Inviting scholars working from different theoretical perspectives and with different methodologies to think about regulation in this regard, we hoped that the collection would help advance a conversation about an important concept in communication studies. While the chapters offer many points of convergence or complementary argumentative trajectories, there are also clearly differences among these perspectives. In all cases, our hope is that the collection points to ideas to be examined by researchers working from varied backgrounds who share an interest in coming to a deeper understanding of communicative practices in workplaces and the professions.

REFERENCES

Blyler, N. R., & Thralls, C. (Eds.). (1993). *Professional communication: The social perspective.* Newbury Park, CA: Sage.
Doheny-Farina, S. (1986). Writing in an emerging organization: An ethnographic study. *Written Communication, 3,* 158-185.

Driskell, L. (1989). Understanding the writing context in organizations. In M. Kogen (Ed.), *Writing in the business professions* (pp. 125-145). Urbana, IL: National Council of Teachers of English.

Harrison, T. (1987). Frameworks for the study of writing in organizational contexts. *Written Communication, 4,* 3-23.

Pare, A. (1993). Discourse regulations and the production of knowledge. In R. Spilka (Ed.), *Writing in the workplace: New research perspectives* (pp. 111-123). Carbondale: Southern Illinois University Press.

Porter, J. (1993). The role of law, policy, and ethics in corporate composing: Toward a practical ethics for professional writing. In N. R. Blyler & C. Thralls (Eds.), *Professional communication: The social perspective* (pp. 128-143). Newbury Park, CA: Sage.

Selzer, J. (1983). The composing processes of an engineer. *College Composition and Communication, 34,* 178-187.

PART I

Understanding Regulative Processes, Practices, and Effects

CHAPTER 1

Using Texts to Manage Continuity and Change in an Activity System

Dorothy Winsor

Because texts produce a stable representation of shifting reality, they are tools that heterogeneous groups of people use to regulate one another's actions so they can work together. In this chapter, a study of documents in engineering activity systems shows, however, that people must continually shore up the agreement a text represents because those who wrote the text could not predict everything that might happen and because the tensions that exist between heterogeneous groups make it difficult for participants to agree on what a text meant in the first place. When divergent groups work together, regulation is ongoing because people's different interests will exert disruptive influences on any agreement that the groups have reached. In order to see the ongoing regulation, however, researchers need to look at how a document is used and supplemented over time. The regulatory documents in this study were constructed in their use and not just in their composition.

Rhetoricians have recently begun to use cultural/historical activity theory to examine the way that groups of people coordinate their actions to achieve a common object (Bazerman, 1997; Russell, 1997; Winsor, 1999). An activity system's object is the problem, space, or focus of activity upon which it acts. For instance, a group of computer programmers might have as their object the software that they are writing. The idea of a common object is important because it is one of the means by which an activity system is defined. As Miettinen (1998) points out, "In an activity system, the object of activity has constitutive significance. The object of the activity 'defines' the activity" (p. 424). In other words, an activity system exists only if people within it are acting upon a common object. However, the theory also assumes that most activity systems are heterogeneous, made of up of people with differing backgrounds and interests (Engstrom, 1992). Thus, common objects would be achievements, not naturally occurring situations. Computer programmers, for instance, can be divided in how they define their task or can take up projects that their managers have not assigned (Artemeva & Freedman, 2001).

Moreover, although activity theory is most commonly represented spatially by a set of triangles, it actually assumes that activities occur over time, or as Russell (1997) says, as they are historically shaped. The theory is commonly used to examine activity as a verb not a noun, a trait it shares with the sociology of Bourdieu (1977, 1990), who argues for the importance of the fact that social relationships exist not just synchronically as structures, but in time. Because social relationships function in time, they almost inevitably change. One interesting problem that activity theory invites us to consider, then, is how a heterogeneous assembly of people can agree upon a common object and act in concert over time. How do the members of disparate communities regulate their behavior to invite at least temporary cooperation and coordination? From the perspective of rhetoric, the process is bound to be a rhetorical one, but we have little data on how the process might actually work.

In this chapter, I wish to argue that because they produce a stable representation of shifting reality, texts are among the tools used both to create common objects and to coordinate activity over time. Generic texts are particularly useful for creating common objects because they not only use a typified form but also invoke a typified action. My claim extends McCarthy's (1991) observation that groups sometimes create what she called "charter" documents intended to "stabiliz[e] a particular reality and se[t] the terms for future discussions" (p. 359). Such charter texts are actually rather common, ranging from a society's sacred texts, such as the U.S. Constitution or the Bible, to the more mundane texts that I propose to discuss in this chapter: a Request for Quote (RFQ), which serves as an agreement between a supplier of goods or services and a client company, and a labor contract that regulates the relationship between management and blue-collar workers in an automobile plant.

An RFQ and a labor contract are particularly interesting documents because they draw together groups of people who might be expected to have difficulty reaching and maintaining agreement. When McCarthy and Gerring (1994) examined the revision of the *Diagnostic and Statistical Manual of Mental Disorders* (DSM) of the American Psychiatric Association, they found that tension was created by the fact that mental-health professionals came from various schools of thought, but all of them needed to be able to use it as a standard in their diagnoses of psychiatric illnesses. However, these professionals are a less obviously disparate group than are a buyer and seller or management and labor. Buyers and sellers as well as managers and workers are normally taken to have opposing interests, rather than just varied ones. Yet they must interact in a joint activity system requiring a common object. Such interaction necessarily involves a certain amount of regulation, but as I will demonstrate, the regulation occurs in a way we might not expect, a way that enacts Foucault's (1980) claim that power is relational, and that we all participate in enacting it. As he says, "power is in reality an open, more-or-less coordinated (in the event, no doubt, ill-coordinated) cluster of relations" (p. 199). The texts I will examine function in historically shaped circumstances to manage both continuity and change over time in joint action among disparate groups. In so doing, they are resources that groups use to try to regulate the actions of others so that their own goals are attained.

If the RFQ and the labor contract that I examine here are typical, these charter documents exhibit two characteristics. First, they are most often communally written because a communal writing process is one means by which people calibrate their perceptions and actions so that they are in harmony. That is, the generic conventions of these documents require that people negotiate in writing them. In turn, the need for negotiation suggests at least some level of shared power among participants—a sharing that is enacted partly by the generic expectations for the text. While a formal power hierarchy might nominally exist, groups who write charter documents together are acknowledging that, at least at the point of the document's creation, no one group is in a position to impose its standards and desires on the others by fiat. As Law (1994) observed in his examination of the management of a scientific lab, "ordering is certainly not the preserve of those who give the order, though it may be sometimes the latter's wish that this were so" (p. 3). The charter document is thus intended to define what group members can expect from one another and the rules by which they will each operate. In effect, it is meant to establish a regulatory system in what would otherwise be an uncoordinated relationship. Those groups who are in a position to participate in writing the charter document can also participate in structuring those regulations.

The second characteristic of these documents is that they usually exist in the middle of a genre system made up of the central stable text and subsidiary reinterpretations. Such subsidiary interpretations are necessary because, as the evidence presented here suggests, charter documents can never be fully successful in articulating agreement both because writers can never articulate all aspects of a situation and because situations change. Moreover, because groups are disparate, we might expect that tensions would exist from the creation of the document that would constantly stress it. Nonetheless, such documents must continue to function if the activity system is to continue on its current track, or indeed even if it is to take a new one that needs to be coordinated.

Thus, imperfection in these documents cannot invalidate them, but must instead be repaired. Eventually this repair might be done through revision, as with amendments to a constitution or communally negotiated changes to the DSM

(McCarthy & Gerring, 1994). However, documents such as a labor agreement or an RFQ often function like legal contracts and so are closed to revision for some previously agreed-upon period of time. Indeed, as with the patents that Bazerman (1994) examines, their legal status is one of the sociotechnical resources that writers bring into play to give these documents weight and hence use them to achieve some degree of stability amid the normal flux of reality. Legal force is one of the ways that, as Law (1994) says, activity systems "try to convert verbs into nouns" (p. 103). Thus instead of being revised, these documents must often be reinterpreted through subsidiary texts like judicial opinions or the multiple texts that help practitioners use the DSM (Berkenkotter, 2001, p. 340). This reinterpretation occurs amidst a certain amount of ontological sleight of hand, however. For it is usually reached by pointing to the original document and claiming that the reinterpretation is what was meant in the first place. If all groups agree, however reluctantly, that this is so, then the reinterpretation becomes what the original document says and disappears as an action on its own, and the appearance of continuity is achieved. The combination of a central stable text and subsidiary reinterpretations serves as a way to simultaneously manage continuity and change in joint activity. The charter document tries to achieve a stable vision of unstable reality; indeed that is its reason for being. But such stability is at best a temporary accomplishment. Therefore the document must constantly be adjusted to that ongoing and not wholly predictable reality. Order turns out to be hard to maintain.

Coordinating people's activities is not a neutral effect, however. McCarthy and Gerring (1994) say that charter documents exist for the "social and political effects" they are meant to generate (p. 147). When disparate groups create the charter document, they naturally try to craft the vision of reality that makes their individual object easier to obtain. The negotiation by which they are written is not free of self-interest, nor does it occur in a world that is free of hierarchy. Regulatory authority is going to be vested in the charter document, and groups want it to represent their interests. Moreover, the subsidiary texts often try to interpret the central text in the best interests of one group or another by claiming the authority of the central one. Control of the subsidiary texts is often crucial in allowing one group or another to exercise power.

In their constant need for reinterpretation, these documents constitute what Bazerman (1994) has called "non-reducible genres" (p. 90), or genres that must be constantly reread and interacted with. Bazerman says that such genres constitute a "fundamental puzzle for genre theory" (p. 90). Genres embody typified textual responses to typified social situations and hence constitute a stabilizing force. But nonreducible genres are typified only if readers continue to perceive them as such during extended interaction. Thus any understanding of genre requires us to know how groups of people regulate their perceptions so that they continue to see these genres as representing a stable reality even in the midst of changing circumstances.

In the remainder of this chapter, I will first describe the procedures used in gathering the data for my discussions. Then I will talk about the process of creating the charter document and the equally important process of maintaining its usefulness through subsidiary texts. The discussion will aim to demonstrate the local and improvisational nature of regulation; the fragility of a common, stable understanding of reality; and the role that writing plays in stabilizing activity through defining a common object. This stabilization of texts into genres that form common objects is a much more active process than we might assume, requiring continued improvisation in the form of subsidiary texts that shape and reshape mutual understanding of the central text, often to the benefit of one group or another.

METHODOLOGY

The observations reported here are part of an ongoing, long-term study of four engineers, which began in 1989 when the four men enrolled in the cooperative-engineering college where I taught. I have conducted interviews with the participants each year since then, asking them to describe their understanding of the writing they were doing at work. (Previous results from this study have been reported in Winsor, 1990, 1996, 1999.) This chapter draws upon the interviews conducted from 1998 through 2002 with three of the participants, Al, Chris, and Ted.

Participants

In this period, which took place between three and eight years after they completed engineering school, the four men were gradually moving into positions of greater responsibility in their workplaces. In 1998, Al was a labor representative in an automobile plant. His job was mediating the interaction between blue-collar workers and management. By 2002, he had been promoted twice and was the Superintendent of Labor Relations, managing employment issues for both salaried and union labor and supervising the work of at least eight people, including some who supervised others. In 1998, Chris and Ted were applications engineers for companies that supplied parts to the automotive companies. They mediated the interaction between their companies and the big firms that bought their companies' parts. By 2002, Chris had gotten an MBA, quit his job with an automotive supplier, and started his own company as a manufacturer's representative. Essentially, he was a salesperson with whom suppliers contracted to sell their technical products to other companies. Ted had completed an MS in engineering, left the supplier company, and was managing large projects for one of the automobile companies to whom his former company supplied parts. Thus, during the period of this study, the participants gradually became more responsible for organizing a set of activities and seeing that they continued to run smoothly. Also, relevant for this chapter, all of their jobs seemed to put them in a position where they had to coordinate the activities of others who would have divergent interests.

Texts

In this chapter, I will examine two charter texts and the subsidiary texts associated with them. The two charter texts are the Request for Quote (RFQ) that defined the tasks that suppliers agreed to do for automotive companies and thus regulated the worlds in which Chris and Ted functioned, and the labor contract which Al called the "Bible" of his workplace.

The RFQ

I am using the term RFQ to stand for three genres that are interlocked: the Request for Quote, the responding Quote, and the resulting Purchase Order. Taken together, these genres regulate the central activities of Chris's and Ted's work. When a company does not wish to do a certain kind of work itself, it will issue a request for another company, called a supplier or vendor, to bid for a contract. The RFQ spells out the required work in some detail so that the supplier will know exactly what it needs to do and therefore how much to charge. Responding to the RFQ, a supplier submits a bid called a Quote. If the client accepts the Quote, it will issue a Purchase Order and buy the goods or services the supplier offers. This process is utterly mundane. In 1998, Chris estimated that 70% of the writing he did at a supplier was associated with what he called this "cycle of life." In 1999, he said that he wrote 5 quotes a week in a slow week and 20 to 30 in a busy one. Such work was so common that bids were submitted on forms, showing that the generic aspects of this work had been institutionalized by the supplier. In theory, these texts defined work that could extend over several years. In practice, as I will show, they needed to be interpreted by a variety of subsidiary documents if they were to be effective over time.

The Labor Contract

In a manufacturing setting, a labor contract is a centrally important document. The contract is negotiated both nationally and locally, with the national contract between employer and union forming the base upon which additions are created to meet local circumstances at any individual site or division. Al said that the purpose of the contract is to "define how we're going to operate, what are the rules. Without giving up management's right to run the business, we want to put some limitations on what we're going to do, how we're going to operate." The contract is a delicate business, then, seeking to regulate while acknowledging the need to set limits on that regulation. The national contract is beyond Al's jurisdiction, but he does have input into his company's local agreement, and interestingly enough, his employer finds the stable reality of a contract to be too confining. Thus, Al's employer has what is called a "living agreement," meaning that management and labor agree to negotiate the solution to local problems as they arise, rather than wait for the current contract to expire. Any such negotiations, however, have to result in a solution that can be argued to be consistent with the national contract.

This chapter draws on all comments the three participants made about the RFQ or the labor contract between 1998 and 2002 as well as examples of these documents shown to me. In examining these data, I looked for common patterns in the three men's understanding of how these documents worked and what function they served.

COMMUNAL CREATION OF CHARTER TEXTS

I begin by describing the process through which RFQs and labor contracts are created because the process is part of the way that these texts are positioned to promote a common object and cooperative activity among groups with different or even competing interests.

RFQ

In my description of the RFQ, I said that it was issued by a company that wished to hire a supplier to produce goods or services for it. However, that description hides the degree to which the creation of the RFQ is a negotiated process both within the companies involved and between them. For example, in 1999, Ted began a major task for an automotive company: designing and installing a new production line for manufacturing cylinder heads. As a part of this task, he issued an RFQ that called for a supplier to bid on making the new equipment that would go into the factory. Ted said that the 24-page RFQ was the most important document he worked with because it "describes in detail all the features and all of the options that I want with the machine, the terms and conditions, how I want it quoted, and how I want the project management to be handled."

Given the importance of this document, it is perhaps not surprising that pre-paring it took some amount of negotiation within Ted's company. For instance, his draft had to be passed through three levels of management. As Ted said, "Just creating the document isn't the hard part. The hard part is going through several meetings, communicating, and determining content." The involvement of so many layers of management does suggest the degree to which the RFQ is a regulatory document. Managers do not leave its creation to the engineers who will execute it and who know the most about the machinery it describes; rather, they use the occasion of writing the RFQ to negotiate a temporary order in the organization itself and in its relation to the outside supplier organization. Thus we see that while negotiation is in play here, such negotiation does not create the world from scratch in some sort of Habermasian ideal speech situation. Rather, there are sociohistorical resources, such as previously established organizational hierarchy, that some people are able to use in arguing for their particular version of a document and the stabilization of reality that it represents.

More surprising than the internal negotiation, perhaps, was the external negotiation that Ted conducted with the supplier. The supplier was one his company had contracted to use repeatedly and was the only supplier expected to bid. Negotiation was necessary to insure that he was asking for things the supplier could do. So we also see that if the RFQ is to be successful, no one in Ted's company is in a position to unilaterally declare how it will be written. Stability is an effect that one can't generate by oneself in a disparate group such as that constituted by buyers and sellers.

Negotiation was necessary within the supplying company, just as it was within the manufacturer and between the manufacturer and supplier. Chris explained some of the more obscure details in 1999. In preparing a Quote, one of the things he needed to do was ask his company's own factories and engineering sites to estimate costs and negotiate internally to reach a cost that was competitive and profitable. These internal negotiations could be complicated. For instance, Chris said that if he wanted to bring in a large order for his company, he needed to negotiate with the factories to dedicate some of their capacity to this order when they also had other obligations. Also, he knew that his company's factories disliked providing replacement parts for items they had sold previously and thus might be generating high estimates for costs in the area that he would need to encourage them to lower. He said that he dealt with "10 different foundries, and I might as well be dealing with 10 different companies because they can all do their own thing and march to their own drum." That is, even within a company, different groups with their own interests are nearly always involved. In preparing his company's bid, Chris had to achieve agreement through negotiation because he did not have the authority to order anyone to do anything. Chris put it this way: "In the line of work we do, you absolutely, positively will get nowhere if you don't know how to motivate people to do something that they don't want to do and end up not hating you for it in the process." Note then that Chris's comments suggest that agreement must be reached, but that not everyone will be equally pleased with the results. Indeed, the impossibility of pleasing everyone equally is probably one of the reasons that the agreement is temporary. From its inception, it is subject to stresses and strains due to competing visions of the object of their activity.

Also note that this negotiation is usually oral because, Chris said, writing things down made them more "official." The movement from oral negotiation to written document is a significant one. "Every time I put something down on paper, I am setting precedents," he said. Indeed, setting precedents, making public the regulatory regime that all now agree to, seems to be part of the function of the document (cf. Winsor, 1999). Thus writing was held off until agreement was reached. The RFQ/Quote/PO chain of documents was communally written to draw together disparate groups. In these documents, money and engineering effort, supplier and client, were pulled together to define how everyone involved would achieve a common object: the new production line. In these documents, different visions of the object were smoothed over and brought into alignment in a public statement meant to regulate the activity of a number of groups of people with different and sometimes competing interests.

Labor Contract

The process of writing a labor contract almost always begins with the existing contract that has been in effect for a period of years and is about to expire. Both management and union officers identify areas in the contract with which they are dissatisfied and come to the table with proposals to amend it. The need to start with the old contract marks the process of writing a labor contract as somewhat different from the RFQ process. The RFQ probably includes some boilerplate text, but does not depend on past language in the way that a labor contract does. Everyone in a labor negotiation is cautious about changing text because they cannot predict the long-term effect that a change might have. In effect, managers and union representatives function like members of a deeply traditional society in which a text is passed down from generation to generation. The contract, says Al, "is our law. It's our Bible." However, in tension with this traditionalism is the unpredictability of the future.

The contract is negotiated between the two sides, each of whom recognize that they sometimes leave things vague and will have to negotiate further later. As Al said, "You write out what you agree on, and you say anything that's not encompassed in the agreement will be negotiated between the parties." At the point when the contract is written, the two sides are cooperating, but they are also aware that they represent different interests. "You get as far as you can, and then the next shot around, you try to go further," commented Al, showing that he knew that what served management's interests might not serve those of the union members. "I work for management, so my directive is to accomplish the goals that management wants. We try to have an open relationship and be fair and up front, but at the same time, we have to run a business as efficiently as we can." The level of different interests that this statement shows means that from the start, tensions are built into the contract that are probably even stronger than those in the RFQ.

Like the engineers negotiating the RFQs, Al and his union counterparts represent other groups. Al's work is cycled through managers, and union representatives know that they have to get approval from their membership. Also like engineers negotiating the RFQs, labor representatives frequently wish to keep negotiations oral, turning to writing only once agreement has been reached. Writing was reserved for the rules they had agreed to. As Al said, sometimes he did not even take notes: "We're talking about it right now. Nobody wants to be held to anything." Writing was taken to represent something permanent and the end of negotiation, not something to be used prematurely.

THE NEED FOR SUBSIDIARY DOCUMENTS

The RFQ and the labor contract, then, work to stabilize an agreed-upon version of reality so that different groups can work together. However, such stability is a precarious condition that is unlikely to continue without the people involved exerting constant effort to maintain it. In other words, the stability of the central document was not a natural occurrence but an achievement that people reached by means of subsidiary documents that tried to fit contingent events into the world the charter document had described. Another way to put this is to say that participants used the subsidiary documents to try to control what the original document meant. That is, participants tried to write about contingency in a way that would link the new document to the old one and borrow its authority. Each side typically maintained that the interpretation it was putting forth was the one intended by the original document—that it was maintaining an order that the other side was disrupting. And as in the creation of the original document, writers again tried to work for their own advantage.

RFQ

As is evident from the discussion so far, the RFQ-genre system coordinates the activities of many people, including multiple employees of both supplier and client companies. The need for this coordination is one of the reasons that the process of producing these documents takes so much internal as well as external negotiation. However, what is not evident from what I have said so far (and in McCarthy's 1991 discussion of charter documents) is the continual repair that is needed to keep things going after the purchase order is issued. An RFQ may look like a short-term use document, one of Bazerman's "reducible genres" (p. 90): a company issues it, suppliers bid, a quote is accepted, and work starts, with the RFQ's goal accomplished. But the RFQ actually has a long-term function because it is what everyone consults when there is a dispute about who should be doing what for what sum of money. In Ted's case, he and his colleagues lived by the RFQ for the year-and-a-half the line was being built and the subsequent year during which problems with the line were being worked out. During this time, constant interpretive work was necessary to maintain a common object for their work.

The most frequent problem was that two parties might differ on the meaning of the way something was specified in the RFQ or perhaps find that something had not been specified at all because, as Ted said, the writer had assumed that "everyone knew" what a phrase might mean. He said that the hard part of writing these documents was deciding what needed to be articulated versus what was "understood." A situation in which "everybody understands" arises only in a united activity system—a rare occurrence—and not one that flourished in Ted's work.

Ted and his supplier dealt with their differences by resorting to improvisational supplementary texts consisting mostly of letters and e-mails through which they negotiated about whether the supplier's work had met the terms of the RFQ. These texts were surprisingly common. "Most of my e-mails that are directed to my suppliers are to complain about the service and support or the equipment supplied," Ted told me. However, he also believed that he could not coerce his supplier but needed the supplier's cooperation. He could not simply invoke a rule to regulate the supplier's behavior. He shaped his documents to achieve the supplier's cooperation in two ways. First, he tried to link his texts to the original document for their justification and call upon that document's authority, a tactic that those who disagreed with him used as well. Second, he adjusted the tone of his messages to match the degree of proximity he shared with the recipient. He always tried to use objective information as the basis for his arguments, but if he did not know the recipient personally—a condition that usually coincided with the person being farther up the supplier's corporate hierarchy—then he was particularly careful to stay away from emotional language.

For example, in 2001, Ted showed me a letter he wrote to a supplier about a problem with the supplier's machine. In the letter, he referred his readers to the RFQ, told them about the problem, and asked them to see the truth of his claim for themselves. He seemed to assume that once his supplier had looked at the RFQ, agreement would be forthcoming, and he saw no reason to fix the meaning of the RFQ himself. In this case, however, the supplier replied, arguing that they were in compliance with the RFQ and saying that the problem was only "a basic misunderstanding" of what constituted compliance. The supplier, too, claimed to be in agreement with the RFQ. When this happened, Ted had to use other rhetorical resources to make his case. The resource he referred to most often when talking with me was the shared disciplinary value that engineers from any company place on facts.

Ideally, Ted believed that his written message should describe events or facts that were "objective"; that is, that everyone could agree on with minimal discussion. "The way I see it," he said, "as long as you can put it plain as day in front of them and show them exactly that these are the items you missed . . . they can't deny it." The importance of writing in this process is shown by the fact that Ted referred to writing about facts as "documenting" them. To document something, (to put it into a document), was to give it a weight and presence it would otherwise lack, but only if the documentation was skillfully done.

For Ted, skilled documentation was, for the most part, writing that avoided emotional language: "I like to be known as somebody who can write well as opposed to somebody who likes to vent." The emphasis on unemotional language reflected both the engineering valuation of logic and the need to get others to cooperate in sharing his vision of reality. In other words, it reflected a kind of rhetorical awareness of the values that he and his audience shared and also of the need to avoid alienating his audience in a less than unified system. "One thing I think I've done a good job of is I don't think I write emotionally like some people do," he told me. "I think, as a result, that I maintain a pretty good relationship with just about everybody I work with." Ted tried to maintain good relations with people even when there was conflict. Interestingly, the need to maintain good relations suggests that not all emotions are to be avoided, only negative ones. For instance, Ted said he tried to get people involved from the start in planning a project because then they would be more willing to "sign the job" for him: meaning, willing to sign off on what he had done. Thus the positive emotions that come from being consulted and having one's opinions valued could be used to draw people from disparate groups together into a more unified system.

Moreover, Ted was willing to use even negative emotions with people with whom he worked closely because he had personal relationships with those people that would make them want to come to agreement with him. For instance, he said that he usually communicated first with the engineer from the supplier who was on site, and "there may be a lot of subjective comments" in that communication. "I use it to motivate them," he said. But if the people on site could not fix the problem, then he wrote to their superiors, and in that document, he stayed "as objective as I possibly can."

So everyone involved in the RFQ system recognized the authority of the original document as an important resource for arguing their own case. A skilled writer like Ted tried to marshal that resource by getting his readers to accept his interpretation of the charter document. He did so by drawing on the values they had in common—respect for facts and positive personal relations—to persuade them to accept his account of what the original document meant. In some way, it is possible to say that while the meaning of the original document is called on to support the subsidiary ones, the original document sometimes has no meaning until the subsidiary documents establish one. The meaning that people writing the original document thought it had collapses under the strain of contingent events and must be reestablished. A stable document, like a stable reality, is a necessary illusion, but an illusion nonetheless.

Labor Contract

Like the RFQ, the labor contract required subsidiary documents to maintain its effectiveness. Indeed, creating these subsidiary documents was a major part of Al's job, requiring far more of his time than the periodic negotiation of the contract itself. He attributed the need for these documents to the unpredictability of the future: "In labor relations, we have contracts; but things change, so they have to be updated; and every day's a new day." These subsidiary documents had official standing in Al's workplace because they became part of what his facility called its "living agreement." They were officially acknowledged interpretations and modifications of the local contract. On the other hand, as Al said, "I didn't write the national agreement, so I can't change it." Thus he had to be very careful, treading the line between treating the national agreement as fixed and dealing with unpredictable local reality.

In Al's work, the subsidiary documents tended to be more formally recognized than the e-mails and letters with which Ted worked. For instance, a prime example of a subsidiary document is something called a Memo of Understanding, a document that was written when some local action needed to be clarified as consistent with the national contract. One Memo of Understanding that Al showed me concerned the restructuring of a department. Because the national contract specified what would happen when workers were transferred between departments, this particular memo carefully spelled out whether a new department was being formed or an old one was simply being rearranged. Memos of Under-standing were signed by representatives of both management and labor and supplemented the local living agreement. Their purpose was to take unexpected events and describe them in a way that allowed them to be fit into the common understanding that the charter text of the national contract had created.

In creating these documents, Al and everyone who worked for him made a great effort not to introduce new language if they could help it. "Every time you say it a little differently, you could change the meaning of it," he said. "And you don't want to." McCarthy and Gerring (1994) point out that in revising the DSM, any changes were perceived as "potentially disruptive" and thus "task-force leaders had to avoid changes, additions, and deletions—the usual objectives for revising a document—whenever possible" (p. 149). Al faced a similar situation, although his task was even more difficult because he was not supposed to be revising the contract in any way. His concern was that in creating new language to deal with contingent events, he might inadvertently create a "loophole" that the union would use to upset the balance of power that management thought they had created in the contract. So these subsidiary documents derived as much as possible from the central charter one, right down to repeating exact language, even when writers knew the language was ungrammatical or unclear. Al needed to represent his communications as being directly from the text that management and union representatives had all agreed on, unless he wanted to open the door for new problems. In so doing, he helped to maintain the illusion of the central text's stability and therefore helped to maintain the ability of managers and union employees to work together.

The repetition of unclear language could also help to manage the tension between stability and change in another way. Because the language was unclear, it was more likely to require interpretation. For instance, the exact language of the contract was used in most official communications with union members. When I asked Al if people sometimes had trouble understanding that language, he acknowledged that they did, but saw that as a good thing, saying that he expected people to come to him for an oral interpretation of the difficult written text. Al did the interpretation orally because, although the discussion so far has shown that the meaning of any document is never completely fixed, he didn't want to be held permanently to whatever interpretation he was giving. Putting his interpretation in writing would make that kind of fixing of meaning more possible. In other words, surprisingly enough, he expected to vary the interpretation: "It's always subject to interpretation. So sometimes you want to be the able to interpret it differently depending on what's going on in the business." The written language of the contract stayed the same, giving the illusion of stability, but its meaning varied as Al varied his interpretations to meet different situations. The written word in the form of the exact language of the contract served as a fixed point to which people could refer; the spoken word was used to interpret it, and Al wanted that interpretation to remain in his purview. In so doing, he could shape the regulatory force of the contract to management's advantage.

Managers (and presumably union representatives) actually built these opportunities for interpretation into many of their texts. For instance, subsidiary documents were sometimes deliberately left "vague, open to interpretation," said Al. He pointed to a sentence in one Memo of Understanding and said,

> You always want to say that "the parties agree to discuss any problems which may arise as a result of this memorandum." Meaning that we know we put some wiggle room in here and what we've said is that we agree to sit back down and talk about those things that aren't covered in here.

The desire for "wiggle room" showed that Al and his colleagues recognized that reality was not predictable. Law (1994) has argued that managers aspire to only a modest ordering of their organizations because they know that totalitarian control is not possible (p. 40). Within organizations, managers often know that hierarchy is less effective than outsiders sometimes believe. The managers at Al's facility seemed to recognize that any order they managed to achieve was temporary. Things could slip out of their control. The wiggle room in the Memo of Understanding was intended to give them an opening through which to reestablish order when that happened.

Sometimes, however, managers at Al's workplace claimed a level of authority in establishing regulations. For instance, a second kind of supplementary text, called an Administrative Guide, did not become part of the local contract, but it used language from the contract to try to clarify a point. Al showed me one that explained the classification of "team leader." The fact that this document was labeled "Administrative Guide" was meant to indicate that it was not part of the central text and did not require union approval. That is, this document was explicitly treated as an interpretation that management had the right to put forward unilaterally. Of course, the union might disagree that such a right existed, but over the years, labor representatives like Al had established precedents for what fell into their territory, and such precedents were very powerful rhetorical resources. Anyone could use precedents as a rhetorical resource to bolster their own interpretation of the contract.

On the other hand, Al told me that he would not issue an Administrative Guide without checking orally with the union representative and asking, "Do you see anything here that might get you in a bind?" But, he added, "I'm not asking for approval. I don't need approval." Such consultation came about because like Ted, who was trying to maintain good relationships with his supplier when negotiating differences, Al too was concerned about good relationships. "When you're dealing with the union, a lot of times your main focus is on relationship building. . . . In labor relations, all you have is your credibility." Thus, oral negotiation was a way to bridge the gap between management's claim of authority and the need to maintain the common activity system. The oral interaction propped up the subsidiary text and made it part of the common understanding despite its seemingly unilateral origin.

Therefore, as with the RFQ, the labor contract is a central, seemingly stable artifact that managers and union members use to coordinate their activities by creating a common object. However, the participants in this activity system have to work to maintain that illusion of stability and at the same time, try to read the contract so that it helps them achieve their individual, disparate goals. The tactics they use include using contractual language to suggest continuity and negotiating if they have to in order to maintain a common understanding of that language. Such a course of action results in memos of understanding and administrative guides, which help to supplement the contract. However, management at least would like to avoid negotiating if it can and keep the control of the contract's meaning in its own hands. Al did this by using the agreed-upon written language to indicate continuity, while orally interpreting the language to the company's advantage. As a result, writing is almost always treated as a resource for stabilizing changing reality, preferably by regulating everyone's actions to meet the goals of those doing the writing.

CONCLUSION

One of the things that this study suggests is that the term "regulation" may be misleading, at least as it is commonly used. It often seems to imply that people's behavior is regulated when an outside force coerces certain kinds of behavior. This view of regulation ignores the question of how regulation is generated. It also seems to assume that large, preexisting structures create an order that absorbs everyone and thus an order in which genres naturally occur because all members of the order can easily identify typified actions. Activity theory encourages us to ask where regulation (and genre) comes from, how it occurs, and how it is maintained. In this study, what seemed to happen was that participants engaged in rhetorical activity to define what regulations consisted of. They used rhetorical and social resources to try to shape written regulatory regimes to serve their own interests. And while some of these resources were more available to some people than to others, they all participated in the generation of power to the extent that they could participate in the writing.

This study suggests that when an activity system consists of divergent groups, a situation that is actually the norm, regulation is ongoing because people's divergent interests will exert disruptive influences on any agreement that the groups have reached. In order to see the interaction of ongoing regulation and charter documents, however, researchers need to look at how a document is used and supplemented over time. The regulatory documents in this study were constructed in their use and not just in their composition. Indeed, they are still being constructed because a stable document is an elusive goal, just as a stable order is.

This is not to say that local instantiations of regulatory regimes are freely undertaken with no impact from larger orders. While regulation is local and improvisational, Foucault (1980) tells us that local orders get annexed, used, and colonized to make big structures that then affect what one can do locally. However, as activity theory urges us to concentrate on how power is generated locally, Foucault too advocates a bottom-up examination of power's construction. In order to understand the operation of power, he says, we need to start with local manifestations and establish

> the manner in which they are invested and annexed by more global phenomena and the subtle fashion in which more general powers 'or economic interests are able to engage with these technologies that are at once both relatively autonomous of power and act as its infinitesimal elements. (p. 99)

Looking at regulation from this point of view suggests that the kind of tension between disparate groups that I have been describing is a good thing. It keeps us from falling into what Law (1994) calls "the hegemonic belief that we have specified the character of ordering so comprehensively that contingency has been vanquished" (p. 22). If order is locally generated and unstable, not totalizing, then change is always a possibility. We live in a world that is humanly made and that can therefore be unmade and remade. Skilled rhetorical action is one of the resources making such change possible.

REFERENCES

Artemeva, N., & Freedman, A. (2001). "Just the boys playing on computers": An activity theory analysis of differences in the cultures of two engineering firms. *Journal of Business and Technical Communication, 15,* 164-194.

Bazerman, C. (1994). Systems of genres and the enactment of social intentions. In A. Freedman & P. Medway (Eds.), *Genre and the new rhetoric* (pp. 79-101). Bristol, PA: Taylor & Francis.

Bazerman, C. (1997). Discursively structured activities. *Mind, Culture, and Activity, 4*, 296-308.

Berkenkotter, C. (2001). Genre systems at work: DSM-IV and rhetorical recontextualization in psychotherapy paperwork. *Written Communication, 18*, 326-349.

Bourdieu, P. (1977). *Outline of a theory of practice*. Cambridge, MA: Harvard University Press.

Bourdieu, P. (1990). *The logic of practice* (R. Nice, Trans.). Cambridge, UK: Polity. (Original work published 1980.)

Engstrom, Y. (1992). *Interactive expertise: Studies in distributed working intelligence*. Helsinki: Yliopistopaino.

Foucault, M. (1980). *Power/knowledge: Selected interviews and other writings 1972-1977*. C. Gordon (Ed.). New York: Pantheon.

Law, J. (1994). *Organizing modernity*. Oxford, UK: Blackwell.

McCarthy, L. P. (1991). A psychiatrist using DSM-III: The influence of a charter document in psychiatry. In C. Bazerman & J. Paradis (Eds.), *Textual dynamics of the professions: Historical and contemporary studies of writing in professional communities* (pp. 358-378). Madison: University of Wisconsin Press.

McCarthy, L. P., & Gerring, J. P. (1994). Revising psychiatry's charter document DSM-IV. *Written Communication, 11*, 147-192.

Miettinen, R. (1998). Object construction and networks in research work: The case of research on cellulose-degrading enzymes. *Social Studies of Science, 28*, 423-463.

Russell, D. R. (1997). Rethinking genre in school and society: An activity theory analysis. *Written Communication, 14*, 504-554.

Winsor, D. A. (1990). How companies affect the writing of young engineers: Two case studies. *IEEE Transactions on Professional Communication, 33*, 124-129.

Winsor, D. A. (1996). *Writing like an engineer: A rhetorical education*. Mahwah, NJ: Erlbaum.

Winsor, D. A. (1999). Genre and activity systems: The role of documentation in maintaining and changing engineering activity systems. *Written Communication, 16*, 200-224.

CHAPTER 2

Regularized Practices: Genres, Improvisation, and Identity Formation in Health-Care Professions

Catherine F. Schryer, Lorelei Lingard, and Marlee Spafford

Using data from an interdisciplinary study of case presentations conducted by health-care students in medicine and optometry, this chapter argues that genres such as case presentations mediate professional identity formation through providing both regulated and regularized resources. Regulated resources refers to the certain, factual knowledge required in professions; regularized resources refers to the situational, improvisational knowledge and skills that emerge in uncertain practice situations. These two kinds of resources also resonate with the classical debate involved in the meaning of techne. Techne, usually translated as art, referred to both the factual, regulated knowledge that constituted professions such as medicine and the savvy improvisational knowledge and skills needed to handle uncertain situations. In acquiring the resources of their professions through genres such as the case presentations, students are acquiring the "habitus" or the forms of identity and agency associated with their professions with both positive and negative consequences as sometimes tacit attitudes (such as devaluing patient language) are being conveyed that professions might want to question.

VOICES

First Scene: *A third year medical student in a children's hospital is reporting a case to her supervisor. The doctor interrupts:*

Doctor 8: Okay. Now I think, again, a lot of us are taught (pause) if you have inspiratory crackles that's pneumonia.

Student 10: Okay.

Doctor 8: If you have expiratory wheezes, that's asthma. Is that sort of how you thought you were taught (laughs)?

Student 10: Yeah, I guess so.

Doctor 8: Okay. I guess the reality in my experience in these cases is that it tends to be more like a crackling sound. And you may hear some crackish stuff that convinces you it's crackles and wheezes. Um, and that can easily be termed inspiration and expiration, but I think if you saw exclusively inspiratory crackles, that would probably sound a little more like pneumonia. Does that make sense?

Second Scene: *A fourth year optometry student is reporting a case to his supervisor. The doctor comments:*

Doctor 3: The textbooks say you should see them again, but I'm finding I'm inclined to just kind of say come in and see us again if anything new happens. I had one on Sunday. And she noticed the floaters and the flashing lights on Sunday and Monday. The flashing lights have subsided now, but again, I'm inclined to look all around the *fundus,*[1] and if there's no evident breaks, you know, if you don't see that kind of stuff happening, then I'm inclined to say just come and see us if you start seeing flashes, floaters, curtains. But textbook care would be you should see the patient a month later. Then maybe three months later. But that strikes me as overkill, to be honest with you.

INTRODUCTION

In the brief selections above, which are taken from two case studies[2] exploring the socialization of health-care professionals, a medical student and an optometry student are presenting their "cases" to their clinic supervisors. Both students are at

[1] This is a technical term referring to the area at the back of the eye.

[2] This research was supported by the Social Science and Humanities Research Council of Canada, grant #410-2000-1147. We gratefully acknowledge their support as this research could not have been completed without their assistance.

the stage where they are assuming some responsibility for patient care, but both are under the direct mentorship of an experienced practitioner. Our research indicates that this site, which Lave (1993) would call a contested, complex site of situated learning (pp. 8-11), is mediated through what Dias, Freedman, Medway, and Paré (1999) call the "school genre" (p. 44) of the case presentation.

Health-care professionals use case presentations to communicate the salient details of patient cases to one another. On rounds or in clinics, the presenter provides arguments about what ails the patient and how to address this ailment. As is common with institutional genres, the structural features of the oral presentation are standardized and constitute shared knowledge among users of the genre. In the case presentation, data from the interview and physical exam are selected, ordered, interrelated, and emphasized according to two controlling goals: the identification and the treatment of disease.

Besides being central to the activities of health-care, the case presentation also does double duty as an educational tool when students are involved. As an educational vehicle, it has a dual role on the threshold of the health care community. According to Lingard (1998), the case presentation is like "a revolving door: both a method of gate keeping—constraining communicative utterances and sifting out speakers in conflict with community values and goals—and a method of gaining access— generating communication that will succeed in the community and announce the neophyte speaker as kindred" (p. 77).[3]

Prior to their immersion in the work of the teaching hospital or clinic, students are provided with guidelines regarding case presentations. In the hospital setting, medical students are instructed that a case presentation must adhere to the following order: Chief Complaint (CC), History of Present Illness (HPI), Past History, Family History, Social History, Physical Exam, Diagnostic Impression, and Management Plan. According to Mitchell-Ashley, Schryer, Spafford, and Lingard (2003), optometry students, on the other hand, used the visual tool of their clinic's optometric record to structure their presentation.

In our research (Lingard, Schryer, Spafford, & Garwood, 2003a, 2003b; Schryer, Lingard, Spafford, & Garwood, 2003; Schryer, Lingard, & Spafford, 2005; Spafford et al., 2004, 2005), we noted that, as students learn to present cases, they are acquiring not only the *regulated* resources of their profession but also (and more importantly) *regularized* ways of knowing. Both forms of knowing embody professional values and are essential to the process of professional identity formation central to health-care professions. However, the regularized resources are more deeply embedded, more tacit, and far more conducive to the improvisational skills that all practitioners need in order to align their professional knowledge to changing circumstances.

---

[3] See also Lingard and Haber (1999).

In order to explain the salient difference between regulated and regularized resources and the ways they are embodied through the genre of case presentations, we will first focus on the crucial difference between regulated and regularized resources and connect this difference to recent sociocultural theory and to a historic debate in medicine and rhetoric regarding the meaning of art or *techne*. Second, we will discuss how recent theories of genre allow us to argue that "school genres" such as the case presentation mediate the negotiation of regulated and regularized resources and thus facilitate professional identity formation. Finally, we will return to our case studies to illustrate this process and explore both the positive and negative consequences of acquiring these resources in scenes of apprenticeship or situated learning.

REGULATED VERSUS REGULARIZED

The important difference between the meaning of regulated versus regularized is captured in their history of usage. Regulated appears to mean practices that are externally controlled, whereas regularized appears to signal a kind of order that emerges out of a range of diverse practices. For example, according to the online version of *The Oxford Essential Dictionary of the U.S. Military* (2001), a regulated item is "any item whose issue to a user is subject to control by an appropriate authority for reasons that may include cost, scarcity, technical or hazardous nature or operational significance." The online version of *The Oxford Dictionary of Modern Quotations* (2002) quotes William Douglas (2002), a famous American lawyer, as saying that "[f]ree speech is not to be regulated like diseased cattle and impure butter." Objects or concepts that are regulated then are subject to the power and control of external authorities. As it appears in the online version of *The Concise Oxford Dictionary* (2001), this meaning well accords with the meaning of the root verb "to regulate" or "to control or maintain the rate or speed of (a machine or process) or to control or supervise by means of rules and regulations."

The term *regularized,* however, carries different connotations. *The Dictionary of World History* (2000) notes that the expression "hue and cry" referred to

> [t]he practice in medieval England whereby a person could call out loudly for help in pursuing a suspected criminal. All who heard the call were obliged by law to join in the chase. . . . The system was regularized by Edward I in the Statute of Winchester (1285), which rationalized the policing of communities.

The Concise Oxford Companion to the English Language (1998) notes that "the use of printing presses in the Rhineland from the mid-15c promoted standard letters, uniform formats and sizes of paper, and over time more regularized orthographies." This notion of the term *regularize* has several implications: organized practices emerge from agents in particular situations; these practices are more flexible and diverse than the notion of regulation supposes; and agents themselves participate in their own ordering so that regularization does not just proceed from external authorities.

PROFESSIONAL IDENTITY FORMATION

The concept of regularized action, rather than regulated action, accords well with recent theories of socialization and professional identity formation. Structuration theorists such as Bourdieu (1991), Bourdieu and Wacquant (1992), Giddens (1984, 1993), and Engeström (1987, 1993, 1999) have developed theories of socialization that explain how social agents are affected by their social contexts and also how these same agents are not mindless dupes who simply replicate the social forces that help shape their actions.

Bourdieu conceptualizes agents as being embodied in their social practices that they themselves enact in regularized but not totally predictable ways. His insights are particularly useful at explaining how past socialization affects present social action. Two key concepts lie behind Bourdieu's insights: field and habitus. The concept of field or market or game is his way of conceptualizing disciplines, organizations, or social systems. A game, market, or field, according to Thompson (1991), is a "structured space of positions in which the positions and their interrelations are determined by the distributions of different kinds of resources or capital,"—cultural (knowledge), economic (money), social (personal connections) or symbolic (recognition) capital (p. 14). Within fields, agents are struggling to acquire these forms of capital so as to advance their own positions. However, agents also struggle to maintain the position of their fields. For example, traditional medicine endeavors to keep alternative medical practices outside of the realm of its field and deny such practices any form of currency. Certainly during our study we observed health-care students struggling to acquire the ways of speaking and the cultural capital of their field, including Schryer et al. (2003), Lingard et al. (2003a, 2003b), and Spafford, Lingard, Schryer, and Hrynchak (2004).

Bourdieu also effectively conceptualizes agency as habitus. In his introduction to Bourdieu's *Language and Power*, Thompson (1991) explains that for Bourdieu, habitus describes a "set of *dispositions* which incline agents to react in certain ways" (p. 12, emphasis in original). In effect, he further notes, these dispositions "generate practices, perceptions and attitudes which are 'regular' without being consciously co-ordinated by any 'rule'" (p. 12). These dispositions emerge out of social practices such as case presentations. Such tacit dispositions, according to Bourdieu and Wacquant (1992), are "durable" (lasting a lifetime) (p. 13), but they are also generative and transposable or capable of generating a multiplicity of practices and perceptions (p. 126). This set of dispositions endows agents with a "practical sense" or a "sense of the game" (p. 121) of their field.

As Thompson explains, the habitus orients the "actions and inclinations" (p. 13) of agents "without strictly determining them" (p. 13). Most importantly, Bourdieu and Wacquant note that social agents participate actively in this process by investing in their own socialization and their own practices. They assert that "[f]ar from being the automatic product of a mechanical process, the reproduction of social order accomplishes itself only through the strategies and practices via which agents temporalize themselves . . ." (p. 139). Thus agents prepare themselves through their training to reproduce their social settings but never in predictable ways.

Through the case presentation, we observed students being socialized into the ways of seeing, problem solving, and behaving that are characteristic of health-care practice, as indicated by Schryer et al. (2003), Lingard et al. (2003a, 2003b), and Spafford (2004). At the same time, we observed that these students were behaving as agents in definitely strategic ways. They were not simply replicating the structure of the case presentation; rather they were using it as an occasion for regularized improvisations. They were acquiring forms of cultural and symbolic power by their strategic choices, which were always limited by the necessity of adjusting their practices to the expectations of their teachers, the attending physicians. In effect, these strategic improvisations were also shaping their future actions—their habitus—as future physicians who would certainly treat patients, but might also train another generation of physicians.

However, not all these regularized ways of knowing and behaving are in the best interests of these new health-care providers, their future patients, and perhaps even their fields. As Bourdieu (1991) observes, the practices that shape our ability to respond to future situations also prepare us to "mis-recognize" or fail to see the operations of symbolic power (p. 165). He suggests that symbolic power is the "power of constructing reality" through practices, often discursive, that creates a sense of "logical conformism" (p. 165). Practices such as defining, naming, and creating hierarchies construct the commonsense worlds in which we live, and yet they are operations of symbolic power that express deeply held, but often unacknowledged value systems. For example, during case presentations, the students learn to transform the patients' language into medical terminology and, in particular, to observe the medical, commonplace distinction between symptoms and signs. Earlier research by Schryer (1993, 1994) noted the important ontological distinction between symptoms and signs in medical discourse. Symptoms encompass the language that patients use to describe what ails them; signs encompass the complex, intricately categorized language that health-care professionals use to indicate what they believe actually ails the patient. Health-care providers, as Reynolds, Mair, and Fischer (1992) and Atkinson (1988) have shown, believe that the language of signs describes more accurately what the physician sees or witnesses and that the patient language of symptoms has less validity. According to Donnelly (1988), some health-care providers consequently find it difficult to listen to patient accounts and, of course, experience problems conveying diagnostic results back to patients because they lack a shared language.

Bourdieu's concept of the habitus helps explain the kinds of orientations that agents bring to interact with their fields. In contrast, Giddens (1984, 1993) and Engeström (1993, 1999) assist in illuminating how fields or organizations and their tools (symbolic, material) help shape practice. Structuration theorists such as Giddens see agents and social structures as existing in an ongoing, dynamic dialectical relationship and focus on the product of that relationship—social practices. Giddens (1984) observes that "the constitutions of agents and structures are not two independently given sets of phenomena, a dualism, but represent a duality" and that the "structural properties of social systems are both medium and outcome of the practices they recursively organize" (p. 25). For Giddens, social structures, such as already existing workplace practices, shape the behavior of workplace participants.

In our study, for example, medical students were expected to relay their knowledge regarding their patients using the format of the case presentation. As Giddens (1993) makes clear, symbolic structures, such as case presentations, involve both rules and resources that shape agents' behavior. However, rules are not, according to Giddens, absolutely fixed regulations. Quoting Wittgenstein, he says that to know a rule means to "know how to go on, to know how to play according to the rule" (p. 119). Rules provide a sense of structure and also a space for improvising within that structure, a phenomenon that we noted in our study. As noted above, the students had to follow a specified order for the case presentation (rules) or else their mentors would interrupt them and require them to back up and repeat information and analysis. All the students attempted to follow this order; more experienced students, however, had learned how to use the order to their advantage by announcing when they were moving from one section to another or by summarizing the relevant information in a section before moving on. In other words, they were proving to their mentors that they not only knew the rules but that they could use the order strategically to convey their information.

As a symbolic structure, case presentations also contain resources. From Giddens'(1993) perspective, resources, which simultaneously function as constraints, are social practices associated with power. As the following example (Schryer et al., 2003) illustrates, students were required to use medical terminology and not the more common language of their patients.

> Student 5: And she was past 34 weeks. So I guess (pause) she would have been overweight.
>
> Doctor 5: We don't use the term overweight. We tend to talk about LGA.

LGA means Large for Gestational Age. For medical practitioners, LGA simply has more meaning than "overweight," but more importantly, using the term announces the speaker as kindred, a part of the medical establishment—the "we" who use this term. This kind of recognized association, of course, endows the speaker with authority. The constraint that students experience is that they must use this language accurately and correctly or be reprimanded.

At the same time, the case presentation as a set of social practices would not continue to exist unless students and doctors continued to activate it. As Giddens (1993) makes clear, it is by acting as agents, using the organized practices associated with an organization, that practitioners become socialized agents capable of intervening in their social worlds.

Activity theorists, such as Engeström, further illuminate the way organizational practices affect social agents. In its inception in the work of Vygotsky (1978), activity theory emerged as a counterbalance to simplistic notions of socialization that either envisioned individual agents as self-contained, preformed entities (psychological models) or as entities totally at the mercy of their environments (behaviorist model). Instead, Vygotsky envisioned agents as learning by using tools in purposeful, goal-directed activities. He saw that these tools, both physical (hammers, pencils) and cultural (language), preexist their users and mediate the interaction between agents and their social environments. By using tools, human agents internalize the values, practices, and beliefs associated with their social worlds. At the same time as they become experienced users, agents can, in the midst of deliberate activity, affect their social contexts or even modify their tools. Certainly, in our research, we saw that by using the mediating tool of case presentations, health-care students were internalizing the values and practices of medicine while involved in purposeful activities that would lead to their own ability to affect future social contexts; that is, their ability to deal with their own future patients.

Engeström (1987, 1993, 1999) and other researchers (Cole, 1999; Scribner, 1985; Wertsch, 1981) have extended Vygotsky's work into a model for the analysis of complex interactions between agents and social structures in professional and workplace settings. While retaining the concepts of tools mediating the socialization of agents, they have expanded the analytical concepts within the notion of system to account for more of the dialectical, or rather dialogical, interactions that occur between social agents and between social agents and their settings. Engeström (1993) defines an activity system as a system "that incorporates both the object-oriented productive aspect and the person-oriented communicative aspect of human conduct," and he suggests that a human activity system, "always contains the subsystems of production, distribution, exchange and consumption" (p. 67).

Furthermore, activity-system researchers (Capper, 1999; Engeström, 1999; Kuutti, 1999) have developed interesting approaches to help account for change and the ways that agents themselves, after they have internalized their social tools, can affect their social settings. Most workplace settings are characterized by multiple, and even overlapping, activity settings. As participants in those systems, agents can, and often do, bring rules and resources from one system into another and in this way can introduce change or innovation into a system.

Recent work in identity formation has profited from the work of Bourdieu as well as structuration and activity theorists. Based on their empirical studies of groups such as Alcoholics Anonymous and psychiatric patients, Holland, Lachicotte, Skinner, and Cain (1998) have observed that individual and group identities are not static constructs; rather, they note that "identities are improvised—in the flow of activity within specific situations—from the cultural resources at hand" (p. 4). Their metaphor of improvisation captures accurately the tension between structure and agency so well articulated by Giddens (1993). Holland et al. go on to explain that "[p]ersons and . . . groups are caught up in the tensions between past histories that have settled in them and the present discourses and images that attract them or somehow impinge upon them" (p. 4). As they negotiate this tension between their past orientations or habitus and the rules and resources in their present situations, agents "self-fashion" themselves and construct identities as "hard-won standpoints" that are "dependent upon social support" and "vulnerable to change" and yet allow for a "modicum of self-direction" (p. 4). The work of Holland et al. provides researchers with a way to conceptualize the complex relations between social structures and individual actions within workplace settings. Their work offers researchers a way to explain how past socialization affects present situations—although never in totally predictable ways—and how the rules and resources of social settings also affect agents, but again, not in totally predictable ways. In effect, their work adds to our notion that through tools and symbolic structures such as the case presentation, agents are not being regulated so much as they are actively participating in the regularization of their own practices.

GENRE THEORY

Current research in genre theory resonates with this sense of regularized practices. Earlier work tended to focus on the constraining or regulated practices associated with workplaces. Beginning with Miller's (1984) insight that genres coordinate forms of social actions, the rhetorical and North American genre schools have demonstrated the way social and contextual structures shape the ways writers and readers in organizations cope. Paré and Smart (1994) and Smart (1993), for example, have examined the way existing genres shape the reading strategies of banking executives. Winsor's (1996) work on the discourse of engineering, Bazerman's (1988) study of scientific discourse, and McCarthy's (1991) exploration of psychiatric record keeping all illustrate the way existing discourse practices constrain the choices of professionalized writers.

However, more recent work demonstrates the improvisational and regularized nature of genres. Building on previous work in genre and activity systems theory, Bazerman (1994) has observed that genres are, in fact, parts of interrelated systems that connect the past to the future. So for example, in our study, medical students transformed past information (gathered from interviewing patients and consulting written patient records) into a present event (case presentation) that will have future consequences (adjusted records, consultation letters, and the student's grade). Bazerman (2002), like Bourdieu (1991), also notes that participation in these genre events is identity or habitus forming. As Bazerman observes, "genre shapes intentions, motives, expectations, attention, perception, affect, and interpretive frame" (p. 14). In his major studies, Bazerman (1988, 1999) focuses on how well-placed, expert users (Sir Isaac Newton, Thomas Edison) develop a self-conscious sense of agency as they manipulate the genre systems that shape them.

Russell (1997a, 1997b) focuses on the difficult balancing act that newcomers face as they appropriate what are, to them, new genres. Using activity and genre research, Russell (1997b) suggests genres "help mediate the actions of individuals with others in collectives (activity systems)" (p. 514). Russell demonstrates the double bind (p. 533) that students face. Professional identities and forms of agencies are woven into genres, such as the case presentation; and yet, to write or speak those genres, students must eschew other ways of speaking that might seem more comfortable or familiar to them. As Russell (1997a) acknowledges, "Agency is distributed in streams of activity as participants appropriate voices in the networks of disciplinary practice" (p. 230). However, Russell (1997b) points out that genres are not monolithic structures, and they are not totally predictive. Rather, quoting Schryer (1993), he notes these mediating tools are only "stabilized-for-now" (p. 522) and in fact, are reproduced by their users through a range of operations. Because they are involved in multiple activity systems, users can transfer resources from one system to another.

Schryer's endeavors in genre theory have worked toward this sense of agents participating in regularized practices mediated by genres that provide them with the constraints and resources they need to improvise their activities. In her early work, Schryer (1993) argued that genres such as medical records are "stabilized-for-now or stabilized-enough sites of social and ideological action" (p. 200). This definition attempted to capture the tension that she observed as health-care practitioners (veterinarians) used an existing discursive structure and yet adjusted that structure to changing circumstances. In a later study of letters refusing long-term disability benefits (Schryer, 2000), she focused on documenting the range of strategies that writers used to enact this genre in their workplace setting, an insurance company. Based on her findings and echoing Bourdieu, she redefined genres as "constellations of regulated, improvisational strategies triggered by the interaction between individual socialization, or 'habitus' and an organization, or 'field'" (p. 450). The current studies (Lingard et al., 2003a, 2003b; Schryer et al., 2003, 2005; Spafford et al., 2004, 2005) recognize that genres orchestrate not only regulated but also regularized, improvisational strategies. In other words, the studies suggest that power is not only imposed from above or regulated but that agents themselves self-regulate, but not in totally predictable ways. Social agents regularize themselves. Thus, genres are constellations of regulated and regularized improvisational strategies triggered by the interaction between individual socialization, or habitus, and an organization or field.

TECHNE

This distinction between regulated and regularized forms of strategic action also connects to a current debate occurring in health care and in rhetoric regarding the nature of the kinds of knowledge these fields advocate and their status as an art or a science. Health-care professionals (Atkinson, 1988; Berg, 1997; Gordon, 1988) argue vigorously as to whether their fields should be associated with practices related to unique situations such as patient cases or with more objective, scientifically oriented approaches. The debate, in fact, derives from the meaning of the term *techne* or art; it is a debate that surfaces in the earliest medical and rhetorical treatises. The Hippocratic Oath, for example, declares that medical practice is a techne[4] and Plato (1952) in the *Gorgias* declares that rhetoric is not a techne.[5]

Recent scholarship (Atwill, 1998; Cahn, 1989; Dunne, 1993; Papillion, 1995; Roochnik, 1996) is providing new insights into the nature of this debate. This research indicates that techne had at least two quite different meanings. At one level, techne referred to a teachable set of formulae, techniques, or regulated knowledge. Roochnik (1996) provides a succinct description of this form of techne, which he calls "techne 1":

- The techne must have a determinate subject matter.
- It aims to affect a useful result. . . .
- Its subject matter . . . is a complex conceptual whole that can be analyzed into discrete parts, the recombination of which is clearly delineated by a set of rules.

---

[4] The Hippocratic Oath (1962) specifically refers to medicine as a techne. In part, the English translations of the oath read as follows:

> I swear by Apollo . . . that I will carry out, according to my ability and judgement, this oath and this indenture. To hold my teacher in this art equal to my own parents; . . . to consider his family as my own brothers, and to teach them this art, if they wish it without fee or indenture; to impart precept, oral instruction and all other instruction to my own sons, the sons of my teacher, and to indentured pupils who have taken the physician's oath, but to nobody else. (p. 299)

[5] In the *Gorgias,* Plato asserts that rhetoric is a form of flattery rather than an art or technai "because it aims at pleasure without consideration of what is best" and because it is "unable to render any account of the nature of the methods it applies and so cannot tell the cause of each of them" (p. 25).

- The recombination of its parts does not distort the original whole.
- It achieves maximal precision. Its paradigm is orthography.
- Possession of this techne yields complete mastery of the subject matter.
- It is mechanically teachable (p. 70).

Techne 1 represents Plato's conceptualization of the knowledge possessed by different fields. Commentators such as Nussbaum (1986) argue that Plato was advocating for a techne or methodical science of moral virtue, a field that would use the methodology of dialectic to find absolute principles such as Truth. In consequence, he celebrated fields that seemed to have methodologies that created certainty. Mathematics, for example, was a paradigmatic field for Plato. For many scholars the search for certainty has had tragic consequences. Detienne and Vernant (1978), for example, indicate that the concept of Platonic Truth "has never ceased to haunt Western metaphysical thought" (p. 318). Medical researchers such as Gordon (1988) lament that "[m]edicine is propelled by abstractions which are taken as real" (p. 7).

However, other rhetorical and medical treatises from before and after Plato's period reveal that techne did not just mean mechanically reproducible, regulated knowledge. Treatises such as Isocrates' *Against the Sophists* (1982) and Hippocrates' *Ancient Medicine* (1962) and *The Oath* (1962) indicate that techne meant something like a cunning set of flexible strategies that one could learn in order to intervene in and partially control human or natural events. The term, for example, according to Detienne and Vernant (1978), is associated with Prometheus' skill in stealing fire from the gods and thus with "metis" or "cunning intelligence" (p. 3). The word *kairos,* or appropriate time, as opposed to chronological time, also appears in connection with techne. According to Atwill (1998), kairos is associated with techne 2 as it describes the concept of intervening in events at the best possible moment (p. 57). She observes that kairos or "the opportune moment plays a critical role in the arts of medicine, sea navigation, and rhetoric" and that according to some Hippocratic treatises "every disease can be cured, if you hit upon the right moment (kairos) to apply your remedies" (p. 57).

Both Atwill and Roochnik provide rich descriptions of this form of techne. Atwill (1998) suggests that it "marks a domain of human intervention and invention" that is located in "situation and time" and thus stands in stark contrast to philosophical knowledge (p. 7). Among the characteristics that describe techne 2, Roochnik (1996) includes the following:

- A techne (2) has a determinate but not a rigidly fixed or invariable subject matter. For example, the human body, like wood or lumber, is a unit of epistemic content. . . . Because it is complex . . . it is not . . . fixed or invariable.
- It affects a useful result, for example, health.

- It is reliable, but not totally so. It offers "rules of thumb," rather than rigid rules. It . . . requires appropriate responses to particular occasions. . . .
- It is precise, but does not measure up to the standard provided by mathematics.
- It is certifiable and recognizable by the community, but not infallibly so.
- It is teachable but not infallibly so (p. 52).

In our study of health-care students, we saw that both kinds of techne were being enacted during case presentations. Students were often tested on their regulated resources or their knowledge of "facts" and recognized points of certainty. At the same time, however, they were being offered a model of practice that advocated improvisation or adjusting knowledge to specific situations or even challenging what they considered certain knowledge. These moments of situated practice, we contend, are deeply regularizing. They showed students how to reasonably participate in emerging practices and how to deal with change and uncertainty in ways that their fields would find acceptable. Both ways of knowing and acting, moreover, were deeply implicated in establishing professional values and identity for these students.

THE STUDY[6]

The study itself consisted of a multidisciplinary research program investigating the role of the case presentation in the socialization of students training to be physicians or optometrists.[7] The study explores (1) how novices learn the strategies associated with the situated language practice of case presentation, and (2) how this language acquisition shapes novices' developing habitus or professional identities.

Settings

The first case study was conducted within the context of the third-year pediatric clerkship at an urban teaching hospital. The three-week inpatient component of the clerkship studied involves students in patient-care activities where students function as part of a medical team. Students are responsible for admitting a new patient to the ward every three to four days on their call shifts, for interviewing and examining the patient, and presenting their findings (via case presentations) to the team the next day on rounds. The case presentations usually occur at the nurse's station, in the hallway outside the patient's room, or in a small conference room.

---

[6] Ethics clearance was secured from all levels for this study.

[7] The larger study also investigates case presentations as conducted by social-work students.

The second case study was conducted in an optometry school's on-site clinic. Fourth-year students interview and examine patients, and while the patient is waiting, they leave the room for five to 15 minutes to consult with the attending optometrist and to secure permission for more intensive investigation and treatment if it is deemed necessary. The optometrist/instructor is typically supervising four students at once. The case is presented only to the supervisor, although another student might be present.

Data Collection[8]

For each case study, data were collected in two phases: field observations and interviews. In the hospital setting, 19 oral case presentations involving 11 students, 10 faculty members, and the teaching exchanges related to them were observed, audio recorded, and transcribed. Eleven students and 10 faculty were interviewed; most interview participants overlapped with observational participants. A 45-minute interview script was developed during the analysis of observational data. The script consisted of open-ended questions about the nature and purpose of case presentation in the clerkship. Interviews were transcribed and rendered anonymous.

In the optometry-clinic setting, 31 oral case presentations and the teaching exchanges related to them were observed and audio recorded. Eight optometry students and six faculty optometrists participated in the field observations. The observations were transcribed and rendered anonymous. Four of the optometry students and four of the faculty optometrists from the field observations were also interviewed. The 45-minute interview script reflected trends and issues arising from the observational data. Interviews were transcribed and rendered anonymous.

Data Analysis

Using a modified grounded-theory approach,[9] observation transcripts were individually read by four researchers for emergent themes, and discussions were convened to develop, apply, revise, and confirm a coding structure. One researcher applied the coding structures to the complete data sets using NVivo qualitative data analysis software, returning to the group at regular intervals to report on difficulties or emerging patterns in the computer analysis.

---

[8] See Lingard et al. (2003a, 2003b: Schryer et al. (2003), and Spafford et al. (2003), for a fuller account of methodology.

[9] Our use of grounded theory was modified as we first identified emergent themes but then allowed our theoretical orientations to affect our interpretations of those themes.

Results

Transcript analysis yielded results on two dimensions: evidence of case-presentation moves or strategies and evidence of the teaching and learning of these moves or strategies. Five dominant themes were apparent across these two dimensions, including: Thinking as a Student, Thinking as a Doctor, Strategies of Case Presentation, Teaching Strategies, and Identity Formation. In this chapter we will report only on data related to Identity Formation, as the other results have been reported elsewhere (Lingard et al., 2003a, 2003b; Schryer et al., 2003; Spafford et al., 2004, 2005).

In both studies we explored the ways that the supervisors represented professional identity to students. In the medical study we traced out two distinct ways that the supervising doctors used the pronoun *we* to evoke a sense of becoming part of the medical community. Eight of the participating doctors used the pronoun *we* to refer to what *we* doctors know and what *we* physicians do. In the following instance, the doctor is asking the students about weight gain among infants. She (Doctor 1) asks, "Remember the parameters for that? What do we allow babies to do in the first two weeks? [pause] We allow them to . . ." The students respond with medically known, accepted, and general knowledge about average weight gain among infants. In responding to a case, Doctor 6 explains, "Right. So postinfectious is sort of a broader (term); they used to call it poststrep (inaudible word) nephritis, but we know that other infections can also cause it. So it is more broader to say postinfectious."

Again, this doctor is asserting that certain facts are clearly known and the following doctor declares that certain practices are also clearly acceptable or unacceptable. He (Doctor 7) states, "We don't do a splenectomy for splenic crisis."

This pattern of pronouncing known facts also co-occurs with a testing pattern that we noted throughout the case presentations. Students were often grilled about their factual knowledge as in the following excerpt:

> Doctor 8: Okay, so there—there are the first generations, and the big second generation we use is cepheroxin. And we talked about this the other day. What does cepheroxin do better than?
>
> Student 7: What does it?
>
> Doctor 8: Do better than?
>
> Student 7: Do better?
>
> Doctor 8: Then why did you choose cepheroxin for this boy, rather than, um, ceph [inaudible]
>
> Student 7: It's [pause] the coverage [pause] of [pause] staph?

> Doctor 8: No actually, enceph and cephzolin and keflex have very good, um, good staph coverage.
>
> Student 7: It's h-flu?
>
> Doctor 8: It's h-flu. Right.

In this instance, there is only one factual answer, and the student should know it. This testing pattern, combined with the use of the pronoun *we* to indicate forms of professional knowledge and accepted practices, suggests that students are being inculcated into a set of regulated practices, practices that have a kind of techne-1 type of certainty associated with them.

Sometimes, too, tacit messages or attitudes that the profession itself might want to question were conveyed in these scenes of habitus development. For instance, in the following excerpt the physician is instructing students in how to interact with nursing staff when arranging for ventilation masks for asthmatic patients.

> Doctor 10: Usually what we do is like when we are ordering masks—especially for asthmatics . . . I just worry . . . if they (nursing staff) give them automatically. They don't think about whether or not the patient needs them, they just keep giving them.

We noted above that built into the identity formation of doctors is a deep suspicion of patient language. Here we see evidence that the division between doctors and other health-care practitioners could also be a function of such socialization, and yet, a great deal of current research (Bobby, 1994; Clark, 1997; Fagin, 1992) is calling for team approaches to health care.

However, the pronoun *we* was also used in a different way throughout the transcripts, as in the first scene that begins this chapter. A theme of reflective commentary emerged wherein the physicians addressed issues of handling uncertainty and of improvising according to the circumstances of the case. This theme is particularly striking in the following example:

> Doctor 9: Thoughts about that? Treating? We're often caught when we have a space of time where we have to get our act together and most of the evidence and sort things out. . . . Certainly, what we want to avoid in circumstances like this—where you're caught and you've started treatment. It's not, at this point, putting her at high risk to treat her. She seems to be doing well. I think what we need to do is go back to that and simply come up with a solution we're comfortable with. It's not unusual that we face those decisions. Do we stop an antibiotic or not? You know there's a fair amount of art in deciding how you weigh off not to create disruption, but at the same time to avoid unnecessary treatment.

In this excerpt, the physician, with reference to a specific case, is making clear to students that in practice, regulated ways of knowing do not always apply. As practitioners, they will have to improvise or navigate their way to acceptable solutions. In other words, they will find themselves immersed in emerging practices, practices that are not regulated, practices that they themselves must enact and for which they must assume responsibility.

A variation on this theme of tension between two ways of knowing also emerged in the data from the optometry case study. In this data set, the sense of professional identity was complicated by the presence of competing standards of practice. Our analysis suggested that students were expected to understand and navigate the similarities and differences between official, regulated standards of practice, the institution's own regulations, and clinicians' idiosyncratic methods (Spafford et al., 2005).

Across the transcripts, we found instances where the students were expected to know either government regulations or scientific facts that should govern their practice. In the following example the optometrist is testing the student on her knowledge of government-regulated practices:

> Doctor 1: And, um, according to government regulations, what do you have to do in order to provide a full eye examination to the . . .
>
> Student 1: I think you have to provide all the things, like binocular-vision test, refraction . . .
>
> Doctor 1: No. No, you don't.

For the student these regulations seem to have an air of certainty about them.

However, this sense of regulated practices was particularly challenged when it came to places in the transcripts where the student and his or her instructor discussed pupil dilation (i.e., the administration of a diagnostic drug into the eye that temporarily dilates the pupils to obtain a better view of the internal structures of the eye to diagnose certain health conditions). In certain rare circumstances, dilation can cause blindness. Although the swift administration of appropriate therapeutic drugs can prevent vision loss resulting from errant pupil dilation, the optometry clinic was located in a region where optometrists are only allowed by law to deliver the diagnostic dilating agent but not any therapeutic drugs. So pupil dilation considered inappropriate by an optometrist requires an immediate transfer of the patient to a local hospital for emergency treatment, thus delaying needed care and likely resulting in professional and legal consequences. The official optometry standard of practice set by the region's professional body calls for optometrists to dilate every new patient's eyes while using clinical judgment (i.e., dilate if safe and subsequently dilate when indicated). In response to the official standard, the optometry school requires students to secure the permission of their instructors before dilating, and yet sometimes students, because of time pressures within the clinic and because they have accrued experience, will proceed without permission to dilating the pupils. This decision to proceed without the optometrist's stated permission is also mitigated by what the optometry students quickly learn about their clinicians' differing attitudes toward dilation. The following excerpt graphically reveals the tensions between regulated and regularized forms of practice and knowledge:

> Doctor 4: What are the indications to dilate at this point? [pause] Well he's not diabetic, I guess, so that's yeah. So that's not a risk to him. But, and myopia is, but this is sort of at the low end of the range, where you start to be concerned. . . . Depending on the literature you read, it's usually four to six. Some have gone as low as . . .
>
> Student 8: So are you requesting I don't dilate then?
>
> Doctor 4: No, no, I'm just asking you, what the indications are. And if you feel that it's appropriate to do it at this point.
>
> Student 8: It's probably a good call. We just dilate everyone. (laughs)
>
> Doctor 4: Well, that's probably not a good practice, to just do it arbitrarily. Um, you need indications to do procedures. Um [pause] I think it's probably, you know it's been two years, and he's a moderate . . .
>
> Student 8: Well, what would you do, if it was your patient? Would you dilate him?
>
> Doctor 4: Probably would.
>
> Student 8: Yeah?
>
> Doctor 4: Yeah. But I don't fault not doing it. So I think it's being a shade more aggressive than, than you'd be required to be, but it's, it's not ah, it's not essential. It's one of those, it's one of those where there's, you know, it's borderline indications to, to do it.
>
> Student 8: So, should I go ahead and put drops in then?
>
> Doctor 4: Um, Okay.
>
> Student 8: Well, I don't know what you want me to do. [laughs]
>
> Doctor 4: Yes. Okay means yes. [chuckles]

In this scene of situated learning, the optometrist seems to be trying to demonstrate to the student the kinds of improvisational decision making that have to be made for specific patients. In other words, certainty is not always a feature of scenes of professional practice. Our interview data with students, however, noted that students found these situations intensely frustrating (as is evident in the above transcript) and frequently did see them as evidence of clinical judgment in action. They viewed each clinician as an idiosyncratic practitioner and so spent some time trying to figure out what their mentor wanted. For many students, the notion of regularized practices that are not totally structured wherein they have to make crucial decisions for which they will be responsible is a frightening prospect.

CONCLUSIONS

This research has two related sets of implications for professional writing researchers and instructors as well as for health-care educators and providers. The richer sense of techne suggested by this research problematizes the role of textbooks in any professional writing or composition classroom. Although some sense of certainty regarding rules and conventions might seem necessary, in fact, actual practice always exceeds available formalized techne. In some way, instructors must alert students to the fact that absolute reliance on what seem like "rules" and "conventions" will not serve their interests in the long run. This research also indicates that professional writing researchers and educators need to be especially observant of emerging practices or techne in specific situations. Practitioners do develop their own regularizing practices, and these practices do adjust over time. Researchers need to attend to these fluid practices in order to pass this changing knowledge on to future generations, and educators need to teach students how to attend to change. Finally, this research offers writing researchers and instructors a theorized way to debate the classic division between art and science, a division that has proved pejorative for advocates of applied practices and for those of us invested in teaching communication in an active and applied way.

This research also has implications for health-care educators and practitioners. In our previous research, we saw that students and their clinical teachers were accessing a range of strategies as they together negotiated this event called the case presentation. We noted as well that case presentations were regulated and regularized; that is, that these social actors were enacting both tacitly and overtly some regularly occurring features. Their behavior was being structured. At the same time we observed that these agents were acting strategically and improvisationally. Their choices were never entirely predictable. They were acting as agents within the confines of the resources and constraints of the genre.

In the research reported in this chapter, we are noting that the genre of case presentations also mediates attitudes towards the certainty of knowledge and the validity of certain practices. Some forms of knowledge are conveyed to students as if they were true; and certain practices are advocated as *the way we do things*. Our research also indicates that these sometimes tacit messages (devaluing patient information, disparaging other fields) are being sent in this process of identity/ habitus formation and field construction. However, we also note that practitioners know that in scenes of actual practices, when events are in flux, beliefs about certainty might be misplaced. Practitioners frequently have to make judgments in uncertain situations; they have to improvise and use available resources and constraints to manage uncertainty; and they have to be responsible for their decisions even when they are uncertain as to the results. The genre of case presentation, occurring as it does within a setting of situated learning, afforded practitioners an opportunity to convey this illusive and yet central model of practice, a model of practice that is truly a techne, an integration of both art and science.

ACKNOWLEDGMENTS

We gratefully acknowledge the assistance of the graduate students associated with this project, especially Tracy Mitchell-Ashley and Lara Varpio. They assisted with the data gathering and coding.

REFERENCES

Atkinson, P. (1988). Discourse, descriptions and diagnoses: Reproducing normal medicine. In M. Lock & D. Gordon (Eds.), *Biomedicine examined* (pp. 179-204). Dordrecht, The Netherlands: Kluwer Academic Publishers.

Atwill, J. M. (1998). *Rhetoric reclaimed: Aristotle and the liberal arts tradition*. Ithaca, NY: Cornell University Press.

Bazerman, C. (1988). *Shaping written knowledge: The genre and activity of the experimental article in science*. Madison: University of Wisconsin Press.

Bazerman, C. (1994). Systems of genres and the enactment of social intentions. In A. Freedman & P. Medway (Eds.), *Genre and the new rhetoric* (pp. 79-101). London: Taylor and Francis.

Bazerman, C. (1999). *The languages of Edison's light*. Cambridge, MA: MIT Press.

Bazerman, C. (2002). Genre and identity: Citizenship in the age of the Internet and the age of global capitalism. In R. Coe, L. Lingard, & T. Teslenko (Eds.), *The rhetoric and ideology of genre* (pp. 13-37). Cresskill, NJ: Hampton Press.

Berg, M. (1997). *Rationalizing medical work: Decision-support techniques and medical practices*. Cambridge, MA: MIT Press.

Bobby, E. M. (Ed.). (1994). *Making the team work: Proceedings of the 2nd Congress of Health Professions Educators*. Washington, DC: Association of Academic Health Centers.

Bourdieu, P. (1991). *Language and symbolic power*. (J. B. Thompson, Ed.; G. Raymond & M. Adamson, Trans.). Cambridge, MA: Harvard University Press.

Bourdieu, P., & Wacquant, L. (1992). *An invitation to reflexive sociology.* Chicago, IL: University of Chicago Press.

Cahn, M. (1989). Reading rhetoric rhetorically: Isocrates and the marketing of insight. *Rhetorica, 8*(2), 121-144.

Capper, P. H. (1999, May). *Understanding competence in complex work contexts.* Paper presented at the Symposium on Assessing Against Workplace Standards, Cambridge University, England.

Clark, P. G. (1997). Values in health care professional socialization: Implications for geriatric education in interdisciplinary team work. *Gerontologist, 37,* 441-451.

Cole, M. (1999). Cultural psychology: Some general principles and a concrete example. In Y. Engeström, R. Miettinen, & R-L Punamaki (Eds.), *Perspectives on activity theory* (pp. 87-106). Cambridge, UK: Cambridge University Press.

Dias, P., Freedman, A., Medway, P., & Pare, A. (1999). *Worlds apart: Acting and writing in academic and workplace contexts.* Mahwah, NJ: Erlbaum.

Detienne, M., & Vernant, J. P. (1978). *Cunning intelligence in Greek culture and society* (J. Lloyd, Trans.). Sussex, England: Harvester Press.

Douglas, W. O. (2002). *The Oxford dictionary of modern quotations.* Oxford Reference Online. Retrieved December 30, 2003, from http://www.oxfordreference.com/views/ENTRY.html?subview=Main&entry=t93.e511.

Donnelly, W. J. (1988). Righting the medical record. *Journal of the American Medical Association, 260,* 823-825.

Dunne, J. (1993). *Back to the rough ground: "Phronesis" and "techne" in modern philosophy and in Aristotle.* Notre Dame, IN: University of Notre Dame Press.

Engeström, Y. (1987). *Learning by expanding. An activity-theoretical approach to developmental research.* Helsinki: Orienta-Konsultit.

Engeström, Y. (1993). Developmental studies of work as a testbench of activity theory: The case of primary care medical practice. In S. Chaiklin & L. Lave (Eds.), *Understanding practice: Perspectives on activity and context* (pp. 63-103). Cambridge, UK: Cambridge University Press.

Engeström, Y. (1999) Activity theory and individual social transformation. In Y. Engeström, R. Miettinen, & R-L Punamaki (Eds.), *Perspectives on activity theory* (pp. 19-38). Cambridge, UK: Cambridge University Press.

Fagin, C. M. (1992). Collaboration between nurses and physicians: No longer a choice. *Academic Medicine, 67*(5), 295-303.

Giddens, A. (1984). *The constitution of society: Outline of the theory of structuration.* Berkeley: University of California Press.

Giddens, A. (1993). Problems of action and structure. In P. Cassell (Ed.), *Giddens reader* (pp. 88-175). Stanford, CT: Stanford University Press.

Gordon, D. (1988). Tenacious assumptions in western medicine. In M. Lock & D. Gordon (Eds.), *Biomedicine examined* (pp. 19-56). Dordrecht, The Netherlands: Kluwer Academic Publishers.

Hippocrates. (1962). Ancient medicine. *Hippocrates* (Vol. 1, pp. 1-64). (W. H. S. Jones, Trans.). Cambridge, MA: Harvard University Press.

Hippocrates. (1962). The oath. *Hippocrates* (Vol. 1, pp. 289-297). (W. H. S. Jones, Trans.). Cambridge, MA: Harvard University Press.

Holland, D., Lachicotte, W., Skinner, D., & Cain, C. (1998). *Identity and agency in cultural worlds.* Cambridge, MA: Harvard University Press.

Hue and Cry. (2000). *A dictionary of world history.* Oxford reference online. Retrieved December 30, 2003, from http://www.oxfordreference.com/views/ENTRY.html?subview=Main&entry=t48.e1750.

Isocrates. (1982). Against the sophists. *Isocrates* (Vol. 2, pp. 14-22). (G. Norlin, Trans.). Cambridge, MA: Harvard University Press.

Kuutti, K. (1999). Activity theory, transformation of work, and information systems design. In Y. Engeström, R. Miettinen, & R-L. Punamaki (Eds.), *Perspectives on activity theory* (pp. 360-376). Cambridge, UK: Cambridge University Press.

Lave, J. (1993). The practice of learning. In S. Chaiklin & J. Lave (Eds.), *Understanding practice: Perspectives on activity and context* (pp. 3-32). Cambridge, UK: Cambridge University Press.

Lingard, L. (1998). *The rhetoric of genre as initiation: Socializing the novice physician.* Unpublished doctoral dissertation, Simon Fraser University, Canada.

Lingard, L. A, & Haber, R. J. (1999). What do we mean by "relevance"? A clinical and rhetorical definition with implications for teaching and learning the case-presentation format. *Academic Medicine, 74,* 124-127.

Lingard, L., Garwood, K., Schryer, C. F., & Spafford, M. (2003a). A certain art of uncertainty: Case presentations and the development of professional identity. *Social Science and Medicine, 56,* 603-616.

Lingard, L., Schryer, C. F., Spafford, M., & Garwood, K. (2003b). Talking the talk: School and workplace genre tension in clerkship case presentations. *Medical Education, 37,* 612-620.

McCarthy, L. (1991). A psychiatrist using DSM-III: The influence of a charter document in psychiatry. In C. Bazerman & J. Paradis (Eds.), *Textual dynamics of the professions: Historical and contemporary studies of writing in professional communities* (pp. 358-378). Madison: University of Wisconsin Press.

Miller, C. R. (1984). Genre as social action. *Quarterly Journal of Speech, 70,* 151-167.

Mitchell-Ashley, T. D., Schryer, C. F., Spafford, M., & Lingard, L. (2003, March). *Explicit and implicit messages in optometric records.* Paper presented at the Conference on College Composition and Communication, New York.

Nussbaum, M. C. (1986). *The fragility of goodness.* Cambridge, UK: Cambridge University Press.

Papillion, T. (1995). Isocrates' techne and rhetorical pedagogy. *Rhetoric Society Quarterly, 25,* 149-163.

Paré, A., & Smart, G. (1994). Observing genres in action: Towards a research methodology. In A. Freedman & P. Medway (Eds.), *Genre and the new rhetoric* (pp. 146-154). London: Taylor and Francis.

Plato. (1952). Gorgias. In *Plato* (W. C. Helmbold, Trans.). Indianapolis: Bobbs-Merrill.

Regulate. *The concise Oxford dictionary (2001).* Oxford reference online. Retrieved December 30, 2003, from http://www.oxfordreference.com/views/ENTRY.html?subview=Main&entry=t23.e46988.

REGULARIZED PRACTICES / 43

Regulated item. *The Oxford essential dictionary of the U.S. military.* Oxford reference online. (2001). Retrieved December 30, 2003, from http://www.oxfordreference.com/views/ENTRY.html?subview=Main&entry=t63.e6660.

Regularized. *Concise Oxford companion to the English language* (1998). Oxford reference online. Retrieved December 30, 2003 from http://www.oxfordreference.com/views/ENTRY.html?subview=Main&entry=t29.e1150.

Reynolds, J. F., Mair, D. D., & Fischer, P. (1992). *Writing and reading medical health records.* Newbury Park, CA: Sage.

Roochnik, D. (1996). *Of art and wisdom: Plato's understanding of techne.* University Park: Pennsylvania State University Press.

Russell, D. R. (1997a). Writing and genre in higher education and workplaces: A review of studies that use cultural-historical activity theory. *Mind, Culture and Activity, 4*(4), 224-237.

Russell, D. R. (1997b). Rethinking genre in school and society: An activity theory analysis. *Written Communication, 14*, 504-554.

Schryer, C. F. (1993). Records as genre. *Written Communication, 10*, 200-234.

Schryer, C. F. (1994). The lab versus the clinic: Sites of competing genres. In A. Freedman & P. Medway (Eds.), *Genre and the new rhetoric* (pp. 105-124). London: Taylor and Francis.

Schryer, C. F. (2000). Walking a fine line: Writing "negative news" letters in an insurance company. *Journal of Business and Technical Communication, 14*, 445-497.

Schryer, C. F. (2002). Genre and power: A chronotopic analysis. In R. Coe, L. Lingard, & T. Teslenko (Eds.), *The rhetoric and ideology of genre* (pp. 73-102). Cresskill, NJ: Hampton Press.

Schryer, C. F., Lingard, L., Spafford, M., & Garwood, K. (2003). Structure and agency in medical case presentations. In C. Bazerman & D. Russell (Eds.), *Writing selves/ writing society.* Fort Collins, CO: The WAC Clearinghouse and Mind, Culture and Activity. Retrieved December 1, 2003, from http://wac.colostate.edu/books/selves_society.

Schryer, C. F., Lingard, L., & Spafford, M. (2005). Techne or artful science and the genre of case presentations in healthcare settings. *Communications Monographs, 72*, 234-260.

Scribner, S. (1985). Vygotsky's uses of history. In J. V. Wertch (Ed.), *Culture, communication and cognition: Vygostkian perspectives* (pp. 119-145). New York: Cambridge University Press.

Smart, G. (1993). Genre as community invention. In R. Spilka (Ed.), *Writing in the workplace: New research perspectives* (pp. 124-140). Carbondale: Southern Illinois University Press.

Spafford, M. M., Lingard, L., Schryer, C. F., & Hrynchak, P. K. (2004). Tensions in the field: Teaching standards of practice in optometry case presentations. *Optometry & Vision Science, 81*, 800-806.

Spafford, M. M., Lingard, L., Schryer, C. F., & Hrynchak, P. K. (2005). Teaching the balancing act: Integrating patient and professional agendas in optometry. *Optometric Education, 33*, 21-27.

Thompson, J. B. (Ed.). (1991). Editor's introduction. In P. Bourdieu (Ed.), *Language and symbolic power.* (G. Raymond & M. Adamson, Trans.) (pp. 1-31). Cambridge, MA: Harvard University Press.

Vygotsky, L. S. (1978). *Mind in society: The development of higher psychological processes.* Cambridge, MA: Harvard University Press.

Wertsch, J. V. (Ed.). (1981). *The concept of activity in Soviet psychology.* Armonk, NY: Sharpe.

Winsor, D. (1996). *Writing like an engineer: A rhetorical education.* Mahwah, NJ: Erlbaum.

CHAPTER 3

Who Killed Rex?
Tracing a Message through
Three Kinds of Networks

Clay Spinuzzi

When we turn to sociocultural accounts of human activities, we gain insights into the co-constructed, communitarian nature of those activities, but we often lose a sense of agency, personal responsibility, achievement, and (when warranted) blame. If activity and cognition are distributed across a socio-technical system, so are successes and failures. That leads us to certain questions: Can we blame an *individual* for the failure of a *community's* self-regulative practices? What role do individuals have in implementing and developing such practices? Here, I explore these questions through the untimely death of a dog. I apply activity theory and actor-network theory to the case, exploring how these two sociocultural approaches treat regulative practices—specifically, how they attempt to distribute agency, cognition, and (by extension) responsibility across the sociotechnical network(s). I conclude by suggesting that tracing the genres or following the texts can lead us to further explorations of community and individual self-regulation.

Let's start with the scene of the crime we'll investigate in this chapter. Rex was a dog living in a backyard in "Prairie City, Texas." On a warm summer day in mid-August, Rex's owner called his local telephone company ("Telecorp"[1]) to

---

[1] Telecorp, BigTel, and other proper names in this study are pseudonyms. My 2000–2001 study of Telecorp consisted of observations of 89 workers (from all 20 organizational units of the company) and interviews with 84 of them. I took observations and interviews over 10 months and triangulated them with artifacts collected during the observations.

45

report interrupted local telephone service. A few days later, a telephone service technician opened the gate to investigate the problem. Rex, spurred on by fear or perhaps the prospect of freedom, darted past the startled technician and into the street. Three blocks away, he was struck by a car.

Rex's lifeless body landed in the neighbor's flower garden—and at the periphery of three overlapping networks: a technological network made up of telecommunications equipment; an actor-network composed of a complex assemblage of sociopolitically aligned humans and nonhumans; and an activity network composed of interrelated, constantly developing, often contradictory cultural/ historical activities. Textual knowledge circulates constantly through these networks—but not reliably; somewhere in the thick of the overlapping networks, the vital message "dog in backyard" got stuck. In short, Telecorp's procedures of self-regulation failed; its workers were unable to circulate the message properly through the overlapping networks.

Rex Was Dead. Who Was Responsible?

Legally speaking, Telecorp was responsible and the customer probably ended up suing it. If Telecorp's records were in order, it could turn around and sue BigTel, a major competitor, which strangely enough, happened to be the technician's employer. (On the other hand, if Rex had ended up biting the technician, then blame would have flowed the other way, Rex would almost certainly have been put down, and BigTel would have sued the customer. Small comfort for Rex, for whom fight and flight were equally deadly choices.) But let's leave the legal arguments for the lawyers. Instead, we will conduct this investigation as a wrongful death investigation, following the texts (Callon, Law, & Rip, 1986) or tracing the genres (Spinuzzi, 2003) wherever they lead.

To investigate this death—this catastrophic failure of self-regulation—I will first take a detour by disentangling the three networks, describing them and the insights they separately provide into Rex's death. I will then reentangle the networks and tell Rex's story in a fuller manner, tracing backwards from the moment that I first heard about Rex's death to a few days earlier, when Rex's owner called to complain about service. Once a possible path through the networks has been reconstructed, I will trace forward again. I will level an indictment for Rex's death both on an individual and on the practices of self-regulation that failed her or him. Besides providing some meager closure for Rex's owner, this exercise should provide insight into the question, Can we blame an individual for the failure of a community's self-regulative practices?

SELF-REGULATION

When I talk of self-regulation, I am referring to a notion that takes shape in various ways across the spectrum of sociocultural approaches and that is typically named mediation. For instance, activity theorists conceive of mediation as a way of controlling one's actions from the outside (Vygotsky, 1978, p. 40); when you consult a shopping list to make sure you have the right ingredients for dinner, or you remember a task because you discover a string tied around your finger, you have used an artifact to control or mediate your activity. Distributed cognitionists similarly see mediation as a way to guide one's behavior. Edwin Hutchins (1995), for instance, gives an example of a novice sailor performing a task with the help of both a written artifact and the coaching of a petty officer (p. 312) and elsewhere talks about how historically evolved tools provide a computational ecology that mediates (regulates, guides) sailors' actions. Actor-network theorists go further, roughly equating mediation with transformation (Latour, 1999b) and noting that all elements in an activity network mediate each other; the mediation is that of the entire collective of "actants"—all the humans and nonhumans involved in the activity—and that each new element transforms the entire activity.

These sociocultural approaches make no meaningful distinction between individuals' self-regulation and regulation across a sociotechnical system, or to put it another way, between self-reminders and communication. Both are thoroughly social, both are mediational, and both function to regulate the activity. When I discuss self-regulation, I mean exactly this sense of the term: the material ways in which the sociotechnical system constrains, polices, and sets limits for itself, whether through checklists, e-mail, words repeated subvocally, strings tied around fingers, stories told in the breakroom, or procedures imposed by management. Such self-regulative acts are always social, always meaningful, and belong to the entire activity both in their effects and in their origins. I explore just a few of these self-regulative acts as I discuss the three entangled networks that make up Telecorp.

THREE NETWORKS: AN OVERVIEW

Rex landed at the periphery of three entangled networks. Each network provides very different insights into the events leading up to Rex's death, and it is difficult to sort out how they interrelate. So in this section I take a necessary detour from the main investigation; I disentangle the three networks, identifying and discussing each one. (Remember, I will reentangle them in the next section.)

Network 1: Telecommunications

Telephone companies have service areas, usually designated by city (Prairie City, Texas) or on a map where an entire service area is solidly colored. Rex lived in one of these service areas. But if you take a moment to examine the actual telecommunications network, the physical area it occupies is quite small: copper lines, fiber, switches, repeaters, phone jacks, resistors, inductances, capacitors, transistors, and power sources take up very little space compared to the area they claim to cover. (If you want to see how little space we are talking about, follow your telephone line outside and see where it leads.) As Geoff Bowker (1987) points out in his study of Schlumberger, networks claim large areas, but in practice are vanishingly small; their claim to power is that they transform the world so that things outside the network do not matter (see also Latour, 1987, p. 180; 1993, pp. 117-118). When we think of Telecorp as covering all of Prairie City, we tend not to think of the fact that the network is inaccessible unless we happen to be precisely at the end of a wire with precisely the right equipment. One can imagine that it would be quite difficult to find the end of a wire, since it is so small; but fortunately for us the wires go to predictable places such as office buildings, pay phones, and (most importantly for this narrative) houses. In practice, it is not difficult at all to get on the right end of the right wire, so we tend to forget the overwhelming amount of unwired space and think of the entire city as covered.

It takes a lot to maintain this illusion of omnipresent service, but that work tends to be done quite well. Through diligent regulation of their own and other allies' work, telecommunications companies have turned a thin, fragile, and tentative tangle of filaments into a network that is so ubiquitous and so reliable that, to individual customers, it seems remarkable for service to be interrupted. Yet telecommunications networks are continually being disrupted—by careless backhoes that sever fibers, by power interruptions that scramble servers, by miscommunications with other telecommunications providers, and by hundreds of other events. Workers are constantly being dispatched to repair this fragile and constantly decaying network. Moreover, the network is always being extended further to include new buildings or to provide additional lines. Both repairs and extensions tend to involve ruptures and thus expose the network's shakiness and limitations, but their impact on individual customers is so rare that the network as a whole seems solid as a rock. When individual customers do encounter a rare (to them) rupture in telephone service, they tend to react with shock and outrage.

How extraordinary that the network should appear so solid despite the innumerable disruptions that are actually occurring! In my observations[2] at Telecorp, two cables in its interstate fiber network were severed by a contractor digging up a water main, causing Telecorp to reroute calls over another company's network until the fiber could be repaired. Cross talk on a customer's line

---

[2] This study was designed for breadth rather than depth and for minimal intrusion into work, so observations were limited to 2 two-hour sessions per worker, recorded in detailed field notes, with the second observation followed by a 15-30 minute semistructured interview. In addition, I examined internal surveys, legislation and regulations surrounding the telecommunications industry, telecommunications glossaries, and consumer information.

was traced to a poor cable splice. A switch had to be rebooted because of a citywide power failure. These were everyday events. Customers are continually being inconvenienced—just not the same customers each time. The network is made of exceedingly thin bits of metal, glass, and plastic that have been wired together by a variety of providers, yet it maintains the illusions of ubiquity and near-absolute reliability.

The illusion extends even further. Service itself is an illusion, at least in the sense that it appears to be unitary. For example, Rex's owner has the illusion that his local telephone company is Telecorp, but his phone service is a so-called resale. Although he calls Telecorp with problems, pays his bill to Telecorp, praises Telecorp when problems are fixed quickly, and curses Telecorp when his dog gets struck by a car, his actual service provider is BigTel. Telecorp receives a cut of his payment for being a middleman. As one Telecorp employee succinctly explained to me in an interview, customers pay Telecorp to deal with BigTel so they do not have to.

Even when customers are not resale customers, their phone service is typically a collaboration among providers. Locally, providers hook up to each others' equipment and networks; long-distance providers continually lease space to each other and bargain-hunt for leased space. So even when we are talking about the physical network, we are talking about many copresent, overlapping, spliced networks controlled by various companies. According to Latour (1987), the technical, organizational, disciplinary, political, and economic complexity of the system is black-boxed—that is, hidden behind a simple interface. Customers just see the phone company. The black box is maintained through thousands of daily, localized, typically ad hoc acts of self-regulation: checklists, electronic notes, and immense stacks of printouts inscrutably marked up and highlighted. Such self-regulative acts usually work quite well—but not always.

Network 2: The Actor-Network

That brings us to the second kind of network we will be discussing here. An actor-network is composed of many entities or actants that enter into an alliance in order to satisfy their diverse aims (Callon et al., 1986, p. xvi). Each actant enrolls the others; that is, it finds ways to convince the others to support its own aims. The longer these networks are and the more entities that are enrolled in them, the stronger and more durable they become.

We can easily see telephone service in Prairie City as an actor-network: Telecorp, BigTel, and other service providers have enrolled each other to their mutual advantage, forging alliances and setting up mutually regulating relationships that allow them to maintain the illusion of unified, ubiquitous, and reliable service. But examining these actants gives us only part of the story.

Actor-network theory[3] is a materialist, non-Cartesian approach, and as such it does not draw lines between humans and nonhumans. Telecorp's workers might enroll BigTel's workers, but they also enroll fiber, switches, and pets like Rex, all of which are considered actants; all of which in turn enroll the workers in their own aims; and all of which can turn traitor in any instant. Cables can be cut, switches can fail, and dogs can fight or bolt right out of their yards. Only by lengthening the network—enrolling still other actants—can the network remain durable. Cables are thin and vulnerable in their underground paths, so Telecorp must protect them with signs warning contractors not to dig them up, avenge their damage with fines and lawsuits, and have other contractors at the ready to splice them. Switches depend on uninterrupted power, so Telecorp must provide it to them and quickly restore it if it is cut off. Dogs want to remain safe and are liable to be frightened or acquire sudden wanderlust if they are disturbed, so Telecorp must tend to their needs, asking their owners about them and instructing technicians (techs) about their proper care. When these actants betray Telecorp by being cut, by going down in a power failure, or by escaping from the backyard, Telecorp must attend to these ruptures quickly and reenroll the actants. Everywhere in the actor-network, regulative allies strengthen and lengthen the network, keep actants in line, and reenroll unruly actants so that the company can maintain its black box—the illusion of unitary service. Those allies are lightweight and insubstantial, mere slips of paper and e-mails and folders, but they are powerful actants nonetheless.

This principle of symmetry between humans and nonhumans as actants in a network seems odd at first, and it is tempting to imagine its proponents living in a Disney cartoon with singing teapots and talkative doors. (In fact, the protagonist in Bruno Latour's (1996a) experimental novel *Aramis* spends some time trying to have conversations with doors, keys, and other nonhumans.) But as Susan Leigh Star (1995) argues, the principle of symmetry should not be mistaken for pantheism (pp. 21-22). Rather, it is an attempt to avoid Cartesian dichotomies by applying the same concepts and vocabularies across the entire system. Once that has been done, we begin to see the same sorts of actions taking place whether the actants are human or nonhuman. The actor-network is continually finding ways to strengthen its existing alliances and make new ones; actants are continually convincing their allies to support them in their aims and routing around traitorous actants with the help of other allies. This actor-network, in other words, constantly engages in a project that is essentially political and rhetorical.

---

[3] I am applying the term "actor-network theory" (ANT) broadly. Many of the scholars (Latour, 1996b, 1999a; Law, 1999, 2002) who originated actor-network theory are now claiming to be post-ANT. Yet their post-ANT work looks quite a bit like their ANT work and uses most of the same terms and concepts, so I will use the term "actor-network theory" loosely to describe contributions in this tradition.

But in focusing on the political/rhetorical movements of complex hetero-geneous networks, actor-network theory misses important aspects of the story. In particular, the principle of symmetry leads it away from the study of cognition, human competence (other than the ability to form strategies and alliances), intentionality, and what Reijo Miettinen (1999) calls "learning, development of expertise, complementarity of resources, and know-how in knowledge construc-tion" (p. 182). To recover these parts of the story, I turn to activity networks.

Network 3: The Activity Network

As Yrjo Engeström (1996) points out, activity theory has come late to the notion of networks. Activity theory is equipped to study groups and organi-zations, and indeed its unit of analysis, the activity system, is meant to examine developmental activity of a collective that cyclically works to transform an object. For instance, workers in the Network Control Center work cyclically to transform the Telecorp network; their work involves maintaining, repairing, and extending it. Activity systems tend to develop complex sets of mediational means (including self-mediational or self-regulative artifacts), divisions of labor, and rules as they work their transformations.[4] But the activity system does not scale up well to study broader social phenomena. That is not just happenstance. Activity theory's roots are in the Soviet Union, where, as Kaptelinin (1996) states, "the opportunity to study social phenomena was limited for political reasons" (p. 63). It is hard to explore questions of society and culture freely when simply asking those questions can get you thrown into the gulag. Activity theory's sociopolitical limitations, some charge, led it to represent people as simple executors—the ideal citizens in a system regulated by command socialism (Letorsky, 1999)—and to ignore issues of power and dominance (Häyrynen, 1999). Later activity theorists, particularly in western Europe and North America, have belatedly begun to develop activity theory in ways that account for how discrete activity systems interact, interpenetrate, and coevolve in complexes; that is, in ways that involve questions of power and politics as well as development. These activity networks are composed of activity systems that have become interlinked. This interlinking is often (perhaps too often) described in terms of inputs and outputs (e.g., Korpela, Soriyan, & Olifokunbi, 2000; Korpela, Mursu, & Soriyan, 2002). Activity networks can also become interlinked or interpenetrated because they share a tool, resource, or community (see Helle, 2000; Russell, 1997; Spasser, 2002; Spinuzzi, 2003).

The term *activity network* is consciously modeled after that of *actor-network*; most of the work using this term cites actor-network theory and provides explicit comparisons with it (Engeström, 1996; Engeström & Escalante, 1996; Miettinen, 1999; Miettinen & Hasu, 2002). These two frameworks have much in common, as

---

[4] For introductory texts on activity theory, see Nardi (1996) and Spinuzzi (2003).

Miettinen (1999) points out; they are materialist and monist; they focus on "the concrete networks of actors instead of interrelations between macro- and microscale phenomena"; they draw on distributed resources for doing and acting; and they allow for the independent activity of objects (p. 171). Additionally, as I noted above, they both have a strong interest in mediation and instability. But the two frameworks are nevertheless quite different. Unlike actor-networks, activity networks assume asymmetry, casting nonhumans as mediators rather than actants. They emphasize development, foregrounding human ingenuity, examining learning, highlighting ways that human competence might be distributed among and mediated or regulated by artifacts, and focusing on cultural/ historical development of individuals, activity systems, and the activity networks themselves. And they exhibit structure, both in their nodes (the component activity systems) and their links (the contacts and exchanges between those activity systems).

Let's take these one at a time. Since actor-networks assume symmetry between humans and nonhumans, Latour says, they deemphasize human cognition, volition, and ingenuity (1996b). Like political power, these are seen as traits of an entire network rather than attributable to individuals. In contrast, activity theorists emphasize exactly these traits, which makes sense when you remember that activity theory is based on Marxism with its valorization of workers. Activity-theory's account is unapologetically *asymmetrical*. Cables do not have interests; they are tools meant to mediate human communication or objects to be transformed through repeated actions such as splicing and connecting. Switches do not yearn to continue running; they are materials designed and maintained by collectives of workers. Dogs are unpredictable organisms that can temporarily become the object of an activity network when they escape or bite. In sum, activity theorists tend to see humans as actors and nonhuman artifacts as simply mediators in which human activity has been crystallized or imbedded (cf. Latour, 1996c). An activity theorist would never have a conversation with a door, like the protagonist does in Latour's novel *Aramis,* but she or he would imagine the embedded activity, the dead labor[5] of the workers who designed and manufactured it, which, if you think about it, sounds just as bizarre. The story told by an activity network is one in which human activity and ingenuity play the foregrounded role, even when implemented in a self-regulative tool such as a checklist.

Part of that foregrounded role involves development. Actor-network theory might provide a historical account of actor-networks as they build associations, but it avoids any explicit account of human competence; activity theorists, in contrast, make human competence, development, and learning central points of their studies, as in Engeström (1999, p. 19). You see cultural/historical development everywhere you turn in an activity network. Individual workers are inducted

[5] "Dead labor" is Marx's term; Latour criticizes it in (1996c).

into the network through formal training, informal apprenticeship, trial and error, and socializing. The objects of the component activity systems change. For instance, when Telecorp entered the local telephone market, the entire organization had to adjust to providing a very different service with different practices, technologies, rules, and time scales. The division of labor changes constantly as Telecorp attempts to adjust to offering new products and services. Of course, self-regulative mediational means such as computer programs, documentation, scripts, technologies, and ad hoc innovations are constantly being learned, adopted, adapted, and discarded on both individual and collective bases in attempts to properly regulate the net's work.

Finally, activity networks are much more structured than actor-networks. Engeström (1996) complains that "Latour's actants seem to have no analyzable inner structure; they are like monads or amoebas" and he suggests "stopping to discover the intermediate institutional anatomy of each central actant—that is, the historically accumulated durability, the interactive dynamics, and the inner contradictions of local activity systems" (p. 263).[6] The activity network at Telecorp involves several different activity systems, each developing in different directions under different influences. The activity network involves other organizations besides Telecorp: BigTel, the Texas and U.S. legislatures, contractors, customers, etc. Activity systems both inside and outside Telecorp draw from their fields, trades, and disciplines. Network Control Center personnel generally learn their trade at Telecorp, but they learn to use their tools, pick up vocabulary, and develop shared practices through interactions with their counterparts at BigTel and other telecommunications companies. Activity theorists argue that by ignoring the very different structures in these activity systems, actor-network theory cannot develop a deep understanding of a given network.

At this point, we are tempted to see actor-networks and activity networks as rival frameworks and to do what others (Engeström & Escalante, 1996; Miettinen, 1999) have done—compare them and declare a winner. Instead, I will examine them as entangled networks, switching between them to provide explanations about Telecorp. This is a messy and imprecise approach, but my purpose is to generate many insights rather than to provide theoretical rigor. In doing so, I aim to hold someone accountable in Rex's death.

That may be a difficult task because neither network is especially good at holding individuals responsible. Miettinen (1999) accuses actor-network theory of providing a "managerial" analysis (p. 190), but just look at the climax of Latour's novel about the failure of a mass transportation system, *Aramis*. Who "killed" Aramis? According to Latour's protagonist, the culprits were everyone involved in the project, including himself and his research assistant, even though they were only called in to do a postmortem! What kind of managerial analysis

---

[6] See Engeström and Escalante (1996) for an extended illustration of this point.

does not allow you to hold someone accountable? Actor-networks are filled with traitors, but none can be condemned for their treachery because they are all seen as actants pursuing their own interests; there is no "totalitarian centre" (Latour, 1993, p. 124) by which to judge them, so there is no one with a special authority to demand accountability. Similarly, activity networks are filled with contradictions, cultural/historical tensions that cause actors to oscillate, oppose each other, and change their practices. But such contradictions are seen as happening at the level of cultural/historical activity, as group movements rather than individual mistakes. "Expertise," Engeström (1992) tells us, "resides in collective activity systems" (p. 11) rather than in the heads and hands of individuals. Studies based on activity theory are filled with examples of people who commit errors in their habitual operations and conscious actions and who feel contradictions between interpenetrating activities, but few describe how a person who should know better simply *screws up*. Yet I think—and I am sure Rex's owner would agree—that blaming everyone and no one is a cheap trick. Can we identify the regulative act that failed? Can we find someone to hold accountable in Rex's death?

TRACING BACKWARD

Now that we have discussed the three networks, let us return to the case at hand and see if we can solve the mystery of who killed Rex. We will start at the moment when the news of Rex's death was broken to the workers in the Network Control Center. Who are these workers? Let's look at two descriptions: Telecorp's official description and the description by an NCC worker.

Telecorp Job Description:
[Trouble ticket coordinators are] responsible for opening Trouble Tickets and processing to completion. Responsible for the surveillance of network outages as well as scheduled and/or emergency maintenance notifications. Also responsible for providing notification of emergency and regular scheduled maintenance to other IXC and/or customers. . . .

Eddie's Job Description:
Basically, I analyze a customer's problem, try to identify and correct it to what facilities I have. 'Cause I do have remote access and I do look in the switch and see if it's set up right. . . . I'm here to troubleshoot a customer's call and see if I can get it corrected. Where I think some of the problems I get shouldn't be in my realm, but we end up getting them because every section is overwhelmed.

As you walk into Telecorp's corporate headquarters, the first thing you see is a large set of plate-glass windows looking into a wide semicircular room with high ceilings. On the far wall is an enormous screen showing a computerized map of Telecorp's fiber network. Workers in headphones (mostly young men) face the screen, sitting at long tables in front of computers. The scene is reminiscent of NASA Mission Control. Since the Network Control Center (NCC) is soundproofed, visitors get the impression that it is silent.

Inside, closer to the action, an observer gets a very different picture. Phones ring constantly, workers call to each other or make general announcements, and sometimes the NCC's manager himself calls the attention of the entire NCC to a particular problem. On their screens, workers typically have several windows open and may work on four different tickets (that is, database entries describing customer problems) at once, often typing a summary for one ticket while discussing a completely different ticket with another customer on the phone. Workers send instant messages to each other announcing that they have a particular customer on the line or asking for information on a given job. Notepads and sticky notes are used frequently, though discreetly, since workers are aware that part of their job is to appear neat and orderly to the visitors on the other side of the plate-glass window. Printed lists show the tickets each worker is assigned for that day, and their annotations on these help them keep track of what work they have done. The NCC is a site of intense, often collaborative problem solving with minimal training, sporadic apprenticeship, and a great number of texts.

It was here, as I began my observation of a trouble-ticket coordinator (Donald), that I first heard about Rex. Between calls, a fellow coordinator (Nathaniel) leaned over to Donald and said, "BigTel let somebody's dog out and it got run over. Nothing mentioned in the ticket about it." Donald nodded and listed the tickets he was to work for that day. A few minutes later, the NCC's assistant manager sternly told the story again, this time to the entire NCC. "There was nothing about a dog on the ticket," he said. "You *must* note that." If this were an actual murder mystery, he might as well have announced, "The killer is in this room."

What had happened? I asked Nathaniel, who had taken the angry call from Rex's owner.

> We get down to the level where we ask our customers, "Do you have locked gates or pets?" We had an incident—I came in Monday morning, the first call I got was a man that called me up: "Hey, tech came out this past week to fix my phone. He let my dog out." Dog got out, got run over, lady three blocks away found the dog in her flowerbed. And—see, so—to mention in the notes about whether he had pets or locked gates. OUR tech didn't do it. But [pause] we contract BigTel to do our services. So if that person sues us, that means we have to go to BigTel and sue them for the same thing.

So although early in the mystery, we already get our first big break. Why was a BigTel technician responding to a call for help by a Telecorp customer? Because they *shared* the customer. Telecorp strives to be seen as a single continuous company: customers call in, order service, get service, and get monthly bills, just like the many other utilities to which they subscribe. But recall that Telecorp's service is really a patchwork of different services provided by different teams and even different companies. In this case, Telecorp resells BigTel's service, acting as a buffer for those who do not like and do not want to deal with BigTel. But even when it provides its own service over its own network, Telecorp has to interface with other providers' networks in relationships that are renegotiated monthly. As Telecorp has expanded into other markets, it has associated with more entities and developed more internal departments to deal with them. All of these allies must be continuously coordinated and negotiated with, and each can turn on the others at the drop of a hat. Telecommunications partners are constantly litigating against each other, recording each others' mistakes, and documenting their own procedures in case those records are needed in a lawsuit—as Nathaniel matter-of-factly points out. Most of the time, those ruptures are kept invisible to the customers, who can black-box the whole mess and think of it as the telephone company.

We have established that someone somewhere dropped the ball: a vital phrase, "dog in backyard," somehow became detached from the rest of the message. That person was not the BigTel tech, who could not be blamed because he depended on that phrase to govern his actions. Of course, as Nathaniel points out, BigTel can be held accountable in the legal sense for the tech's carelessness, but we are after someone else—the culprit who failed to pass along that vital bit of information in the first place. So where did the rupture occur? One clue comes from the interview with Donald:

> It sounds like whoever opened the ticket—of course it's, I haven't seen the ticket so I'm not sure myself. A lot of times we don't hear from the customers directly. I mean, they do, certainly that's great. But a lot of the time businesses will call their sales rep, the sales rep will call us—which has some problems sometimes because they may not know all the information that we need to know for a particular problem. Uh, but certainly whenever the customer does call in or if they call to Customer Service, and they don't transfer the call to us, they just say, "Hey, this person called in and I have this information." And I usually ask, "well, do they have any pets or locked gates?" And you hear this dead, this uncomfortable silence and you realize, "Well, no, you didn't ask, did you?" So then you gotta try to call the customer back, and sometimes they may have no line, and if there's no dial tone, there's no way to get ahold of them. They may have been calling from a pay phone or something. So I usually go ahead and put in the notes, "Attempted to contact customer for access information and was unable to contact." . . . And even when I started we didn't ask for pets or locked gates. But a couple of months later something similar happened where a pet got out apparently but they were able to recover it, fortunately [pause] the BigTel tech went and didn't know, and let them out. So they said, go ahead and start asking for local on A2 and A4 tickets, the residential tickets.

Two points jump out at us here. Let us take the second one first. "Even when I started we didn't ask for pets or locked gates," Donald says. Donald has only been working at Telecorp for 12 months. So the practice of asking about pets and locked gates is only about 10 months old. (That is not a very old practice, but then again, high turnover at the NCC means that it predates many of the workers.) Furthermore, it came about because of the experience of some predecessor of Rex's: some luckier dog who managed not to wind up in the neighbor's flowerbed. Let us call this predecessor Fido. Those of you who had doubts that we could trace Rex's death into the thick of the overlapping networks that make up Telecorp, here is your proof. Fido changed the way Telecorp did things, and he did not even have to die for it. Perhaps Rex's death will make that change stick, serving to further regulate how trouble-ticket coordinators do their work, although, as we shall see, that is doubtful.

Now let's turn to the other point. Although the assistant manager had in effect announced that "the killer is in this room," Donald thinks otherwise. The assistant manager may accuse the people under his authority, but Donald claims that these people have internalized the rule of asking about pets and locked gates. (I personally observed few NCC workers asking this question.) Instead, Donald points the finger at the people who usually forwarded the information to the NCC: Sales and Customer Service. That is problematic to Telecorp management because it turns out that the line between customers and workers at Telecorp is extremely porous; at least three departments could have taken the message, so Fido's (and now Rex's) story will have to be repeated to all of them in one form or another. Furthermore, these three departments have especially high turnover, even by the standards of Telecorp, a rapidly growing company.

Things get a bit sticky at this point, since I had only limited access to the trouble tickets and was unable to find out whether the original complaint had been handled by Customer Service or Sales. Customer Service is the most likely, since Sales deals primarily with business customers rather than residential customers. Again, let's look at Telecorp's description of the job, followed by the description given by one of the workers.

Telecorp job description:
[Customer service representatives are] responsible for performing clerical functions in the customer service area, which will include interacting with customers and other carriers on a daily basis.

Margo's Job Description:
My primary job is to answer the phones, make sure to take customer calls, get them whatever they need . . . .

Customer-service clerks sit in low-walled cubicles in a large, open-plan room on the first floor of the corporate headquarters, just across the hall from the NCC.

Unlike the NCC, they are not on display; the casual visitor cannot see into the customer-service department. They have taken advantage of this fact by personalizing their cubicles with holiday cards, stuffed animals, printed anecdotes and jokes, and scriptural references. They are predominately women. Like the trouble-ticket coordinators, the customer-service clerks wear headphones plugged into their telephones and sit in front of computers. They are surrounded by tools: the telephone, which rings almost continuously; the computer, which runs a variety of software; forms; sticky notes; message pads; and the ever-present spiral notebooks that customer-service clerks use to record almost every event. Of all these tools, the message pad would most likely have been used in taking down the information Rex's owner called in. Although customer-service clerks were aided by rudimentary scripts, checklists, and (most importantly) forms, none of these served to remind them about asking about pets and locked gates.

I did not witness the original complaint being filed, of course, but I did observe customer service clerks dealing with similar complaints. None of them asked about pets in the backyard, just as Donald said. None of them demonstrated self-regulative practices or genres that would guide them in asking the crucial question. Are they the culprits? Perhaps not: when they called in trouble tickets to the NCC, they usually kept the customer on the line. According to Donald, he would often ask about pets, and customer-service workers would confess that they had not asked. But the clerks I observed, if asked the question, would be able to simply put the NCC worker on hold and quiz the customer. We cannot rule out customer service entirely, but their culpability seems less likely than Donald suggested.

Perhaps the fault lies with Sales, then? Let us look at how Telecorp describes the position of sales representative (rep) and how one of the sales reps describes his work.

Telecorp job description:
[Sales representatives are] responsible for securing new business accounts in his/her assigned territory. Also responsible for providing after sale and continuing customer service to existing accounts.

Luther's job description:
Day to day, probably a third of my job is taking care of issues of people calling in, problems that they've had, whether that's with their phone lines or things weren't set up properly. Things like that, and most of those, probably I'd say a third of those, are things that there was a glitch somewhere; so I'm having to backtrack, trying to find out—when did that occur, what happened, where does that go next? And a lot of that is still real confusing. Is that a Control Center issue? Is that, y'know, which part of the department—some of that is confusing.

CLEC Sales (that is, *local* sales) operates out of Telecorp's downtown building, some miles from the corporate headquarters where the NCC and Customer

Service are stationed. That physical difference was paralleled by the cultural difference. Workers in CLEC Sales saw themselves as in competition with each other; a large whiteboard displayed each rep's sales for the week, and reps and sales assistants talked about sales in terms of winners and losers.

Donald in NCC had said that customers sometimes left their complaints with Customer Service and sometimes with Sales. How do sales representatives interact with customers? Luther said that about a third of his interactions involve such complaints. But the required action was not clear. "Where does that go next?" he asked. "Is that a Control Center issue?" How should the problem be routed through the overlapping networks? Luther was at a disadvantage because he and two other sales reps had been hired recently and were still undergoing training. But even sales reps with longer histories at Telecorp had similar problems—not because they had not learned the company structure but because the structure kept changing on them. One sales assistant hinted that some of the older sales reps didn't even understand some of the newest services. Neither did the new reps, who had to be trained by technical personnel rather than their colleagues in Sales.

I did not observe any sales reps asking about pets and locked gates. But that is not surprising: sales reps primarily sold to businesses, not residential customers. Businesses offer higher profit margins and, perhaps as importantly, a centralized contact for multiple lines. It is less difficult and more profitable to keep one company rep happy with their 30 lines than it is to keep 30 residential customers happy with their individual lines. It is not surprising, then, that sales reps had developed no self-regulative genres—no scripts, checklists, or forms—to remind them to ask about pets and locked gates.

So can we rule out CLEC Sales as our culprit? Not entirely. Although people in CLEC Sales rarely handle residential calls, they do take on these unprofitable accounts if they are having trouble meeting their quota. And since they have not adopted self-regulative genres for dealing with residential customers, they probably are not asking the right questions.

TRACING FORWARD

Let us summarize, tracing the events forward through the networks instead of backward.

1. Rex's owner notices interrupted phone service.
2. He calls Telecorp to complain about his phone service, speaking to Customer Service (most likely), CLEC Sales (less likely), the Network Control Center (even less likely), or some other department (remotely possible). He does not mention that Rex is in his backyard.

3. A customer-service rep might relay the message to the NCC; a sales rep might call the NCC, Customer Service, or even another department. The message eventually wends its way to the NCC.
4. An NCC worker then calls her or his counterpart at BigTel and sets up a time for a technician to visit the customer's residence.
5. BigTel's technician visits the customer's residence, letting Rex escape.

We still have not managed to pin the blame, but we do notice something in this description. Our attempt to trace forward through the networks becomes bogged down because the three networks all widen and tangle at points 2 and 3. Along which phone lines was the customer's call routed? What actants in the actor-network handled the call and moved it along? Through what component activity systems did the message wend? Even if we were able to determine the answer for this particular case, blaming Rex's death on a specific individual, this snarl in the networks remains. The borders between Telecorp and the customer are too porous, too far-flung to guard them all. Like the hygienists in Latour's (1988) *The Pasteurization of France* who sought to rebuild entire cities hygenically because they could not pinpoint the cause of disease, Telecorp faces an impossible task if it tries to set up safeguards across all capillaries of its actor-network. It cannot inculcate the same procedural or regulative knowledge in every worker in every activity system of its far-flung activity network. That solution is hopeless; but one customer-service worker suggested it anyway in an internal survey dated March 2000. "All products can start or finish in customer service. Every Telecorp employee should know what Telecorp sells and services," she stated confidently. "All employees should sit in a CS rep seat for a day and actually be expected to take care of whatever daily task that day."

This is in fact how Telecorp used to be run. One worker, Karina in Accounts Payable, who had been with the company for 18 years, recalled that in the company's early years, everyone began in Customer Service and "knew everything from the ground up" so "you're not just doing it blind":

> I like to know enough about it that I feel like I know when I look into an account what I'm looking at. That's something we don't do anymore. . . . Setting up an account is not like it used to be, it's a lot harder.

Indeed it is. Telecorp began by reselling long-distance service, but now it offers an ever-expanding number of services: calling cards, long-distance pagers, DSL, Internet dial-up, mobile service, conference calling, and the list continues to grow. Who could learn all aspects of the business? How could Telecorp encourage sustainable practices when the labor shortage made it easy for workers to job-hop, and the booming telecommunications business made it imperative to hire more people—factors that increased Telecorp's turnover in 2000–2001? Telecorp was constantly changing; it was a moving target.

Yet in 2000–2001, Telecorp was attempting a strategy along these very lines. It initiated a new cross-training program in which workers spent time in various sections, learning how the company works. It encouraged workers in each department to apply for jobs in other departments so that each department could share knowledge about the others. One sales assistant (Sheila) was explicitly recruited from Customer Service to help resolve the contradictions between the two departments. (Sheila laughingly recalled that a coworker in Customer Service threatened to come "kick my ass" if she developed an attitude like others in Sales.) Cross-training might help, but as Telecorp becomes more complex, local knowledge has a shorter shelf life. In the few months since Shelia left Customer Service, Telecorp had already split data entry from the workers' other duties—a rather fundamental change in the department's division of labor. How much of Shelia's knowledge would be relevant six months later? Could she continue to bridge the widening gap between Sales and Customer Service, or would her former coworker eventually perceive her to be a traitor and feel compelled to "kick her ass"?

Let us put it another way. When Rex's owner called Telecorp to complain about his phone service, he was able to treat the telephone company as a black box (to use ANT's term), a simple interface for a complex and messy sociotechnical system. "The telephone company" does not do justice to the enormous complexity of the telecommunications industry that his request had to traverse, but he neither knew nor cared about that complexity. His job was simple because Telecorp, for him, was simple. But inside Telecorp, management takes the opposite strategy. Rather than building black boxes for its workers, Telecorp's management is trying to open all of those black boxes at once, turning them into Pandora's boxes (Latour, 1999b), multiplying complexity rather than reducing it. The workers' jobs are complex because their relations to other departments, other activity systems, are complex and getting more complex all the time. That is, no department had simple interfaces to other departments; there was no predictable path for information to take and no reliable way to track it and get it through the networks. Instead of the simple inputs and outputs that some activity theorists have envisioned linking the component activity systems, those systems overlap, blur, and interact in unpredictable and unstable ways. Further, with turnover, acquisitions, cross-training, and constant organizational changes, workers cannot close the black boxes themselves by operationalizing (to use activity theory's term) their routine actions. The developmental activity that would allow for operationalizing—the shared tricks, habits, and lore that make black boxes possible—was difficult to sustain because the activity systems were changing too rapidly, drifting apart too quickly, interpenetrating or associating with divergent fields too frequently, and losing too many individuals.

In short, workers at Telecorp could not consistently regulate their own or each others' communicative practices—not with genres such as checklists or forms, not with stern narratives, and not with procedures in manuals. NCC workers might blame other units for Rex's death, but there is no way to guide interactions, enforce compliance, or hold anyone accountable.

So we have found a huge problem in the network, a set of snarled links that will need to be untangled soon. But we are aiming to hold someone accountable, and looking at points 2 and 3 will get us hopelessly muddled in that quest. We cannot assign blame there because we cannot predict the path that the information will take through the network; if we stop there, we will end up like Latour, blaming everyone in the room including ourselves. No, we must instead look for where the networks narrow. That brings us back to points 1, 4, and 5.

We have already ruled out point 5. We know that the information does not make it all the way to BigTel's tech. Similarly, we can rule out point 1, since the customer cannot be expected to volunteer the fact that Rex lives in his backyard. (Imagine trying to disperse that procedural information across Telecorp's customer base and you will realize that point 1 is actually not a narrow part of the network, but rather the broadest part, as you would expect from the periphery.) We must go back to Telecorp. The only point remaining is point 4: the NCC.

PASSING JUDGMENT

So was the NCC's manager correct when he in effect said that the killer is in this room, and was Donald wrong when he denied the charge? The answer is yes, if by that we mean the proximate cause, the individual who did not successfully follow up on the question of pets and locked gates. As insurmountable as that task seems, since the trouble-ticket coordinators are already terrifically overburdened, we can still argue that this individual caused Rex's death. He was the only one who could have stopped it. He stood at the narrowest part of the network and thus had the best and arguably the only chance to be the gatekeeper. If I had the trouble ticket in hand, I could walk into the NCC and point him out myself.

But the crime was a crime of negligence rather than malice, and the perpetrator can credibly cite mitigating circumstances. "I really did drop the ball," he might say, "and I am sorry, but that is what happens when you try to juggle too many balls at once." It is a good defense. The NCC suffers from the same problems of self-regulation that bedevil the rest of the company. Workers are inducted into an overwhelmingly complex job as quickly as possible, without black boxes to simplify the work, or formal training to help them operationalize the actions they must perform, or systematic regulative genres such as scripts and checklists. They have to learn everything at once, primarily through observation, and keep track of what others do across what seems to be a complex, un-black-boxable company. Of course they complain about the lack of formal training, but in the same breath many technical workers affirm that the best way to learn is through this sort of experience; they are thrown to the wolves but believe that, as one worker in Collections put it, "that's probably the best way I learn." "This is a different kind of industry," another worker told me; yet another claimed that some parts of the job are simply unteachable. Yes, it *is* unteachable as long as networks grow and divide without affording a chance for workers to manage that complexity and change. Without relative stability on which sustainable regulative practices can be founded and formalized, workers can learn only by encountering contingencies and tailoring responses to each one of them. Self-regulative artifacts cannot develop and spread across the company because the work is too ad hoc and mercurial to make them possible on a broad enough scale.

Perhaps this training model was adequate in Telecorp's recent past, but in 2000 it was wholly unrealistic. After all, how was this rule that made it a crime not to ask about pets and locked gates promulgated? Through written instructions in a handbook? The answer is no; it was promulgated through ephemeral stories about Fido and Rex, delivered sternly by a manager to overworked trouble-ticket specialists (half of whom were on the phone and literally typing four notes at once) or through stories circulated among those workers during short breaks as they wait for the next call. As hard as I look, I cannot find any evidence that these stories circulated beyond the NCC.

The culprit's defense, by the way, must have been successful. Unlike me, the NCC's manager had access to the trouble ticket and could determine exactly who had handled Rex's owner's ticket. Yet he did not punish the culprit or even name him. So, regretfully, I have only partly accomplished my goal in this chapter. Yes, I can hold someone accountable, but I have to factor in the mitigating circumstances of three entire networks to do it. Telecorp's rapid growth and rapid turnover had outpaced its training culture. We cannot excuse the culprit's lapse (how could we justify that to poor Rex?), but we can certainly understand it. The worker was simultaneously struggling against a constantly changing, constantly expanding and complexifying technical network; a treacherous, seething, circulating actor-network; and an exploding, developing, dividing activity network.

POSTMORTEM

Well, now that we have solved the mystery and leveled an indictment of sorts, let us do a postmortem—not of Rex, but of the entangling approach that I have followed in this chapter. To do so, let us return to the question I asked before: Can we blame an individual for the failure of a community's self-regulative practices?

My approach of entangling the networks is messy, but it lets us oscillate between two distributed theories. When Fred uses a self-regulative tool such as a checklist or note to regulate his own behavior, we find the two networks highlighting different roles for the tool. Actor-network theory sees the situation symmetrically, as actants that mediate each other and that strike a negotiated political alliance—a "Parliament of Things," to use Latour's (1999b) phrase. Activity theory sees the situation asymmetrically; the tool, which is seen as a crystallization of work or dead labor, allows Fred to mediate or regulate his own actions. Guided by the ghosts of the workers whose operations are embedded in the tools, he learns anew how to be a worker and finds ways to develop his work within the changing activity, in turn embedding his own labor in the changes he makes to his tools. Yet in both perspectives, the tool does a good chunk of Fred's work. So it is hard to pin down blame; in distributing intellectual work across networks, the theories also distribute responsibility and agency, competence and (especially) incompetence. Neither lets us confidently assert, as one of his coworkers did, that "Fred is such a fuck-up."

As this case has shown, maybe that is not such a bad thing. Individual failures of competence can lead us to better understand and assess regulative practices and how they interlock across an activity, while still recognizing individuals' responsibility to participate competently. That is to say, deviance within an activity, whether assessed positively as innovation or negatively as incompetence, is a function of how elements of the activity interact and the degree to which they mutually transform each other. Incompetence is the result of freedom, the loosening of strictures in a network that allows humans and nonhumans to interact in a wider number of ways. Given this insight, it is not surprising that Fred's actions demonstrated innovation as well as incompetence; he developed tools that replaced older ones and that were adopted by some of his coworkers. If we are to agree with his coworker that Fred performs incompetently, we must look at how competence is defined and mediated within the nodes of the networks. What regulative tools and practices support Fred's work, and is Fred incompetent because he misuses them or because he uses them too faithfully? When Fred and his compatriots fashion new tools and practices to continue that work, how do these new elements remediate the old ones, causing this node of the network to behave in new ways? Does this nodal drift appear as incompetence from the vantage point of other nodes? As we begin to answer these questions, we begin to see the value of examining competence and self-regulation in socio-cultural frameworks: deviant acts of incompetence and innovation can spark deliberations, in Latour's (2004) sense, on how to rearrange assemblages of humans and nonhumans, how to regulate those assemblages with greater or fewer mediatory means, and how to reconfigure networks in ways that dynamically adjust for changes in their nodes.

Let me end with a word about entangling networks. By bringing the networks together, entangling them, and oscillating between them, I have leveraged the differences between the frameworks without trying to resolve them. As Engeström (1990) points out, contradictions drive change and development (p. 84); and as Latour (1999a) has noted, contradictions are often dealt with by rerouting around them rather than meeting them head-on (p. 16). That is what I have attempted to do.

Actor-network theory's dictum to follow the texts (Callon et al., 1986) is similar to the dictum to trace the genres (Spinuzzi, 2002, 2003); but the former emphasizes how particular texts are translated or enrolled by other actors, while the latter emphasizes how texts are historically developed and enacted in particular activities. These are complementary views that can be paired productively. My study of Rex's death, I hope, demonstrates a little of the potential in such an approach for examining self-regulation and its failures.

REFERENCES

Bowker, G. (1987). A well ordered reality: Aspects of the development of *Schlumberger*, 1920-39. *Social Studies of Science, 17*, 611-655.
Callon, M., Law, J., & Rip, A. (1986). How to study the force of science. In M. Callon, J. Law, & A. Rip (Eds.), *Mapping the dynamics of science and technology: Sociology of science in the real world* (pp. 3-15). London: Macmillan.
Engeström, Y. (1990). *Learning, working, and imagining: Twelve studies in activity theory*. Helsinki: Orienta-Konsultit Oy.
Engeström, Y. (1992). *Interactive expertise: Studies in distributed working intelligence*. Helsinki: University of Helsinki.
Engeström, Y. (1996). Interobjectivity, ideality, and dialectics. *Mind, Culture, and Activity, 3*(4), 259-265.
Engeström, Y. (1999). Expansive visibilization at work: An activity-theoretical perspective. *Computer Supported Cooperative Work, 8*, 63-93.
Engeström, Y., & Escalante, V. (1996). Mundane tool or object of affection? The rise and fall of the Postal Buddy. In B. A. Nardi (Ed.), *Context and consciousness: Activity theory and human-computer interaction* (pp. 325-374). Cambridge, MA: MIT Press.
Häyrynen, Y.-P. (1999). Collapse, creation, and continuity in Europe: How do people change? In Y. Engeström, R. Miettinen, & R.-L. Punamäki (Eds.), *Perspectives on activity theory* (pp. 115-132). New York: Cambridge University Press.
Helle, M. (2000). Disturbances and contradictions as tools for understanding work in the newsroom. *Scandinavian Journal of Information Systems, 12*, 81-114.
Hutchins, E. (1995). *Cognition in the wild*. Cambridge, MA: MIT Press.
Kaptelinin, V. (1996). Computer-mediated activity: Functional organs in social and developmental contexts. In B. A. Nardi (Ed.), *Context and consciousness: Activity theory and human-computer interaction* (pp. 45-68). Cambridge, MA: MIT Press.
Korpela, M., Mursu, A., & Soriyan, H. (2002). Information systems development as an activity. *Computer Supported Cooperative Work, 11*, 111-128.
Korpela, M., Soriyan, H., & Olifokunbi, H. (2000). Disturbances and contradictions as tools for understanding work in the newsroom. *Scandinavian Journal of Information Systems, 12*, 191-210.
Latour, B. (1987). *Science in action: How to follow scientists and engineers through society*. Philadelphia: Open University Press.
Latour, B. (1988). *The pasteurization of France*. Cambridge, MA: Harvard University Press.
Latour, B. (1993). *We have never been modern*. Cambridge, MA: Harvard University Press.
Latour, B. (1996a). *Aramis, or the love of technology*. Cambridge, MA: MIT Press.
Latour, B. (1996b). On interobjectivity. *Mind, Culture, and Activity, 3*, 228-251.

Latour, B. (1996c). Pursuing the discussion of interobjectivity with a few friends. *Mind, Culture, and Activity, 3,* 266-289.

Latour, B. (1999a). On recalling ANT. In J. Law & J. Hassard (Eds.), *Actor-network theory and after* (pp. 15-25). Oxford: Blackwell.

Latour, B. (1999b). *Pandora's hope: Essays on the reality of science studies.* Cambridge, MA: Harvard University Press.

Latour, B. (2004). *Politics of nature: How to bring the sciences into democracy.* Boston, MA: Harvard University Press.

Law, J. (1999). After ANT: Complexity, naming and topology. In J. Law & J. Hassard (Eds.), *Actor-network theory and after* (pp. 1-14). Oxford: Blackwell.

Law, J. (2002). *Aircraft stories: Decentering the object in technoscience.* Durham, NC: Duke University Press.

Letorsky, V. A. (1999). Activity theory in a new era. In Y. Engeström, R. Miettinen, & R.-L. Punamäki (Eds.), *Perspectives on activity theory* (pp. 65-69). New York: Cambridge University Press.

Miettinen, R. (1999). The riddle of things: Activity theory and actor-network theory as approaches to studying innovations. *Mind, Culture, and Activity, 6,* 170-195.

Miettinen, R., & Hasu, M. (2002). Articulating user needs in collaborative design: Towards an activity-theoretical approach. *Computer Supported Cooperative Work, 11,* 129-151.

Nardi, B. A. (Ed.). (1996). *Context and consciousness: Activity theory and human-computer interaction.* Cambridge, MA: MIT Press.

Russell, D. R. (1997). Rethinking genre in school and society: An activity theory analysis. *Written Communication, 14,* 504-554.

Spasser, M. A. (2002). Realist activity theory for digital library evaluation: Conceptual framework and case study. *Computer Supported Cooperative Work, 11,* 81-110.

Spinuzzi, C. (2002). Toward integrating our research scope: A sociocultural field methodology. *Journal of Business and Technical Communication, 16,* 3-32.

Spinuzzi, C. (2003). *Tracing genres through organizations: A sociocultural approach to information design.* Cambridge, MA: MIT Press.

Star, S. L. (1995). Introduction. In S. L. Star (Ed.), *Ecologies of knowledge: Work and politics in science and technology* (pp. 1-35). Albany: SUNY Press.

Vygotsky, L. S. (1978). *Mind in society: The development of higher psychological processes.* Cambridge, MA: Harvard University Press.

CHAPTER 4

The PowerPoint Presentation and Its Corollaries: How Genres Shape Communicative Action in Organizations

JoAnne Yates and Wanda Orlikowski

We view the PowerPoint presentation, and genres more broadly, as both enabling and constraining human action. We trace the historical development of the business-presentation genre over the last century, examine the influence of the PowerPoint software tool, and consider the evolving enactment of the PowerPoint-presentation genre in a few organizations. Drawing on this analysis, we highlight the emergence of what we refer to as corollary genres that challenge our conventional understandings of genres as being tightly coupled to particular recurrent situations and communicative purposes. Our analysis points to an empirical blurring of genre expectations around conventional discursive practice, suggesting important implications for the nature of workplace communication in contemporary organizations.

In this chapter, we examine how and with what consequences the discursive expectations of a particular genre shape the ongoing work of organizational actors. The genre we focus on has recently become pervasive in multiple spheres of communicative activity (business, education, government, etc.) and is popularly referred to as "the PowerPoint presentation." Virtually everyone who works in an organization today is familiar with the bullets, formats, templates, and clip art that compose the visual representations associated with this genre. This chapter explores how the use of this genre influences the communicative practices of organizational members, and, in particular, how it enables and constrains their discursive choices and actions.

The notion of regulation in communication is a central organizing theme for this volume. The regulation of action always entails a dual influence where activities and outcomes are both facilitated and limited, or in Giddens' (1984) terminology, both enabled and constrained. Because some uses of the term *regulation* seem to imply too strongly a sense of constraint, we will employ Giddens' terminology to highlight that constraint and enablement are not alternatives (a dualism), but two sides of the same coin (a duality). Indeed, we will view the PowerPoint presentation, and genres more broadly, as both enabling and constraining human action.

Like all social structures, the PowerPoint-presentation genre has been shaped by multiple influences over time, and we will focus our attention on two primary ones. The first is the historical business presentation that emerged as a type of business communication in the early years of the twentieth century, and the second is the technological capabilities afforded by computer-based business presentation software, the most widely known of which is the PowerPoint application tool produced by the Microsoft Corporation. In order to understand how the PowerPoint-presentation genre has been shaped over time, it is important to distinguish between the PowerPoint tool (the software used to create the presentation visuals), the PowerPoint texts[1] created through use of the tool, and the PowerPoint-presentation genre as a whole (in which a person presents to a co-present audience using projected PowerPoint texts as visual aids). While these elements are clearly interdependent, distinguishing them analytically allows us to observe variations in the genre and its use over time. It also allows us to consider the implications of the significant shift in genre norms currently underway as PowerPoint texts are increasingly represented in a broad array of media and used in a variety of different contexts.

In the next section, we establish the theoretical basis for our discussion of genre and introduce the notion of corollary genres—variants of an established genre that are enacted in parallel with it. Next, we trace the historical roots of the PowerPoint presentation. After briefly examining the emergence and evolution of the business-presentation genre in corporations during the twentieth century, we discuss the development and use of the software used to produce the visual aids that accompany PowerPoint presentations. We then draw empirically on a number of research studies of specific contemporary organizations (advertising agencies, consulting firms, and high-tech companies) to illustrate how use of the PowerPoint-presentation genre and its attendant corollaries structure ongoing

---

[1] Our use of the term *texts* here is intended broadly to include the various visual, graphic, audio, and video elements that may be created with the PowerPoint tool.

interaction through shaping actors' discursive expectations. We conclude with implications of our analysis for workplace communication.

THEORETICAL PERSPECTIVE

We embed the notion of genre—a socially recognized type of communicative action—within what Giddens (1984) calls a "structurational" perspective. This perspective focuses on the recursive relationship between everyday activities and the social structures that are the medium and outcome of those activities. Central to a structurational perspective is the recognition that social structures do not exist "out there," but are constituted through the ongoing actions of knowledgeable human agents, actions that are shaped, in turn, by the structures. This recursive relationship is what Giddens (1993) refers to as the "duality of structure." Social structures are thus enacted through recurrent human practices. As Giddens puts it, "The production of society is a skilled performance, sustained and 'made to happen' by human beings" (p. 20). For applications of this perspective to organizational contexts, see for example, Barley (1986) and Orlikowski (2000).

We draw on the structurational perspective to understand genre as a social structure that is interpreted and enacted through individuals' ongoing communicative practices. In an organization, typical genres of communication include memos, letters, meetings, expense forms, and reports. As we have argued in Yates and Orlikowski (1992) and Orlikowski and Yates (1994), building on the work of Miller (1984), these genres are socially recognized types of communicative actions that over time become organizing structures through being habitually enacted by organizational members to realize particular social purposes in recurrent situations. Through such enactment, genres become regularized and institutionalized templates that shape members' communicative actions. Such ongoing genre use, in turn, reinforces those genres as distinctive and useful organizing structures for the organization.

As organizing structures, genres shape beliefs and actions, and in doing so, enable and constrain (but do not determine) how organizational members engage in communication. In many instances, actors draw on established genre norms out of habit to guide a particular communicative act (e.g., implicitly using a standard report format to document project progress). In other instances, actors may draw on genre norms more deliberately to accomplish their communicative purpose (e.g., explicitly choosing the informal (and undocumented) genre of phone conversations in order to discuss a confidential matter). Whether used implicitly or explicitly, as Yates, Orlikowski, and Okamura (1999) have shown, genres powerfully influence the discursive norms of organizational interaction. As we have suggested in Yates and Orlikowski (2002) and Yoshioka, Yates, and Orlikowski (2002), these discursive norms may be understood as entailing expectations about the following aspects of communication: purpose, content, form, participants, time, and place. For analytic purposes, we will treat these aspects as distinct; in practice, of course, they are deeply intertwined.

- *Purpose* (why): Most notably, a genre provides expectations about its socially recognized purpose(s). For example, the résumé genre is expected to convey professional (and sometimes personal) information about an individual and, as DeKay (2003) notes, in the employment context it is expected to promote that individual's abilities and experiences in order to secure an interview.
- *Content* (what): A genre also provides expectations about the content of the communication. The résumé genre is expected to contain specific information about an individual's educational credentials and prior employment experience, including the dates and locations of these accomplishments.
- *Participants* (who): A genre carries expectations about the participants involved in the communicative interaction and their roles (e.g., who initiates the genre, and to whom is it addressed). In the case of the résumé genre, it is generated by the individual described in the document and sent to a set of institutions where the individual is submitting an application.
- *Form* (how): A genre provides expectations about its form, including media, structuring devices, and linguistic elements. The résumé genre is typically structured with sections representing different categories for educational background, work experience, and additional interests. It uses relatively formal language and relies on sentence fragments in the form of bullet points to highlight key achievements. It may be generated on paper and distributed via the mail, or may remain as a computer file and be sent as an e-mail attachment, or even posted to a Web site.
- *Time* (when): A genre often entails specific temporal expectations, although these may not be explicitly stated. Résumés are not always dated; yet, because they indicate the timing of an individual's accomplishments, they implicitly reflect the temporal boundedness of the information.
- *Place* (where): A genre also provides location expectations, and these too are not always made explicit. Résumés must be sent to a specific address, whether physical or electronic, and they will be received and considered in a specific place (typically, the organization at which the individual is seeking a position). They also contain references to the locations where the individual has performed his or her work and educational activities.

When agents enact a genre, their interactions with others are structured by the genre's socially recognized and sanctioned expectations around key aspects of the communication: purpose, content, participants, form, time, and place. By implication, genres also provide information about those aspects of communication that are not sanctioned or practiced by the organizational community. The expectations reflected in genres reveal, for example, who is not empowered to initiate or receive certain genres, when or where certain genres may not be enacted, and what content or form is inappropriate for particular genres. In keeping with the above example, résumés conventionally recount the accomplishment of individuals. They are not generally authored by groups or teams and are not effective at conveying information about collective achievements. The expectations associated with résumés thus reinforce an ideology of individualism that prescribes an autobiographical narrative that charts a life in terms of individual successive engagements in sanctioned activities and legitimate institutions. Gaps in this personal timeline are seen as irregular and questionable. When enacted, genres therefore represent forms of what Schryer (2002) calls symbolic power, serving to both enable and constrain types of interaction and modes of engagement. As discussed in Yates and Orlikowski (2002), genres are indicative of what communities do and do not do (purpose), what they do and do not value (content), what different roles members of the community may or may not play (participants), and the conditions (time, place, form) under which interactions should and should not occur.

As enacted social structures, genres change over time (see Yates & Orlikowski, 1992). Indeed, as Hanks (1987, quoted in Schryer, 2002) notes, genres are improvisations, being "produced in the course of linguistic practice and subject to innovation, manipulation, and change" (p. 81). Thus, in their everyday communication, actors may vary (deliberately or inadvertently) how they enact a genre, and if such changes become widely adopted, the shared discursive expectations associated with the genre may be altered to a greater or lesser extent. Less extensive changes in discursive expectations result in adjustments or modifications to a particular genre that do not transform or replace it. For example, in a study of the historical evolution of the employment résumé, DeKay (2003) found that starting in the 1970s the purpose of promoting the candidate's abilities in order to secure a job interview was added to the existing purpose of factually listing a candidate's qualifications. This additional purpose changed expectations around the content as well as the purpose of the résumé genre, but it did not fundamentally alter its form and functioning. More extensive changes in discursive expectations often lead to the emergence of a new genre that is recognizably distinct from the original. For example, Yates (1989) traces the emergence of the memo genre from a series of changes made over time to the established business-letter genre for internal correspondence (e.g., simplified letterhead stationary, adoption of single subject per letter for easier filing, elimination of salutations and closings, etc.).

One particularly interesting genre innovation emerges when actors modify some of the discursive expectations of a particular genre to produce variants that spin off as derivative genres—that is, distinct (albeit related) genres that are enacted alongside the original and that may ultimately evolve into completely separate genres (e.g., the memo as it evolved from the letter). We term these derivatives *corollary genres* and see their emergence as a broadening of the conventional discursive boundaries associated with particular genres. Genre theory suggests that genres are enacted to accomplish particular communicative purposes in response to specific recurrent situations. When texts commonly associated with a certain genre become produced and received in a variety of recurrent situations, the tight coupling of discursive action and situation loosens. The result, as Lemke (1995) shows, is greater variability and flexibility in textual production and consumption, and the generation of new possibilities and challenges for discursive and cultural change.

As we will show below, the business-presentation genre emerged in response to the recurrent requirement to share complex information with multiple people in face-to-face meetings. As the PowerPoint software became widely available, the business-presentation genre evolved into the PowerPoint presentation, a genre that is now dominant in contemporary presentations. Initially, the purpose and recurrent situations of the PowerPoint-presentation genre resembled those of the historical business presentation—to share complex information with multiple people in face-to-face meetings. However, over time and through different uses, PowerPoint texts have been produced and consumed in a wide variety of contexts with different discursive requirements and social purposes (e.g., Web-based slide shows, printed decks distributed in person or by mail, and PDF files sent via e-mail). These additional uses have spawned a number of corollary genres to the PowerPoint-presentation genre, generating opportunities and ambiguities that both enable and constrain the discursive practices of organizational actors.

A BRIEF HISTORY OF THE
BUSINESS-PRESENTATION GENRE

Business presentations with visual aids existed long before personal computers and PowerPoint software. In the early twentieth century, firms were much smaller than they are today and semiformal or formal presentations, especially those with visual aids, were not common. Presentations given at professional- and trade-association conferences were typically read from manuscripts, and within firms informal discussions were more common than presentations. As Yates (1985, 1989) notes, the first visuals to emerge—graphs and charts based on numerical data—grew out of the systematic management movement's emphasis on recording and comparing data about business operations. The first textbook on graphical presentation of data, published in 1914, was Willard C. Brinton's *Graphical Methods for Presenting Facts.* While Brinton's book focused more on graphs for use in documents or as tools for analyzing data than on visual aids for use in presentations, he also mentioned projecting graphs as lantern slides to accompany a talk. He saw graphs as particularly important tools for communicating with management. "In many presentations it is not a question of saving time to the reader but a question of placing the arguments in such form that the results may surely be obtained" (p. 2). Graphs could be used to present large quantities of data compactly and clearly or to convince management of a particular conclusion.

In her study of communication in late nineteenth and early twentieth century firms, Yates (1985, 1989) shows how managers at the DuPont company used graphs and charts to support their presentations. Comparative charts were drawn up in preparation for meetings of the managers of multiple plants, serving as the focus for analytic and problem-solving discussions. DuPont's chart room provides a particularly interesting example of the use of graphs as visual aids for presentation. Sometime between 1919 and 1922 a special chart-viewing room was designed for meetings of the firm's Executive Committee. The room had 350 charts, updated regularly, displaying various aspects of the return-on-investment data for DuPont's multiple divisions. As the committee members deliberated, responsible division managers would be called in to explain any anomalies in the charts (see Lessing, 1950; Piper, 1938). While this use of graphs as visual aids was described as "uniquely DuPont" (Lessing, 1950), it was copied by many of its customers who traveled from all over the country to DuPont's headquarters to learn about it, according to Krell (2001).

By the second half of the twentieth century, presentations with visual aids were more common in business. For example, Connelly's (1958) compilation of communication practices in industry lists a range of media that could be used for visual aids, from motion pictures to blackboards. The list includes slide projectors, overhead transparency projectors (perhaps the most immediate predecessor of computer projection), and opaque projectors, all of which project prepared-in-advance images (p. 170).[2] Similarly, Morris' 1972 book on business communication listed a variety of possible visual aids, including slides, flip charts, and the overhead projector with transparencies (p. 216). Both Morris and Connelly emphasized the need for simple and clear visuals that serve as a support, not a substitute, for the speaker.

By the 1980s, formal business presentations with visual aids were commonly used to communicate information and arguments to an audience co-present in the same physical space as the presenter. As portrayed in textbooks of the era, such as Robbins (1985) and Munter (1982), overhead transparencies were the visual aids of choice in most internal business presentations, although presentations to large external audiences often used 35 mm slides for a more polished effect. In the decade between 1975 and 1985, according to Parker (2001), the number of

---

[2] One source, Parker (2001), claims that overheads did not "fully enter business life until the mid-seventies" (p. 78), when developments in transparency film made it possible to photocopy directly onto the transparencies. Nevertheless, the presence of the overhead projector in the list of visual aids presented in these earlier texts suggests that this method of presenting visual aids enjoyed at least limited use in business somewhat earlier.

overhead projectors sold per year in the United States more than doubled, from around 50,000 to over 120,000 (p. 78). A speaker could hand-draw the transparencies of this era, which Brooks (2004) notes were often called slides by analogy to 35 mm slides. Increasingly, however, secretaries or designers prepared them with typed or Letraset text (generally all capital letters), drawings, or graphs, photocopied onto the transparency (see Figure 1). Instructors generated some norms for such presentations, including many exhortations not to let the visual aids upstage the presenter (as they often did in a 35 mm-slide show for which the lights were typically turned off) and to make them simple and readable. Both portrait and landscape layouts for visual aids were possible, but with text visuals especially, portrait layout was apparently used more often (e.g., see Munter 1982, pp. 94-97). Bulleted lists with indented subcategories became a common format. Presenters frequently used two techniques for revealing information gradually: covering part of the transparency with paper and sliding that paper down as needed (a practice that audience members frequently found annoying) and using overlays to add information gradually to an image.

"STURDY BATTLER"
(ACCEPT TOUGH, DENY TENDER)

THE BEST WORLD IS ONE OF COMPETITION, CONFLICT, ASSERTIVENESS, POWER

FUNCTIONS IN GROUP TO TAKE CHARGE, INITIATIVE, PRESS FOR RESULTS, DISCIPLINE, STRUCTURE

EVALUATES OTHERS IN TERMS OF WHO IS WINNING OR LOSING, WHO HAS POWER

INFLUENCES BY WILL POWER, ORDERS AND COMMANDS, DOMINATION, THREAT, CHALLENGE

FEARS LOSS OF POWER, BECOMING SOFT OR SENTIMENTAL, BECOMING DEPENDENT

UNDER STRESS MOVES FAST AND TAKES LEADERSHIP (OVERACTIVE AND EXPLOITATIVE)

Figure 1. Typical transparency created using a typewriter.

Note: Adapted from an unpublished overhead transparency created for classroom use by J. Van Maanen, based on content in E. Schein (1969).

By the late 1980s and into the 1990s, personal computers began to play a significant role in the creation of visual aids. Word processing programs could be used in place of typewriters to create text visuals or labels on graphs in different sizes and fonts, as illustrated, for example, in Munter (1992, pp. 111-112). Spreadsheet programs could create graphs of data, as illustrated by Holcombe and Stein (1990, p. 82), and early graphics programs such as Harvard Graphics and PowerPoint itself were beginning to appear. Many of these texts briefly considered color use in visual aids (as in Munter, 1992, p. 108), and a few, such as White's (1996) *Color for impact,* focused extensively on the topic. In the late 1980s, the idea of projecting directly from a computer first emerged, though authors of texts such as Donnet (1988, pp. 55-60) generally cautioned against the dangers of depending on such technology.

In 1987, PowerPoint 1.0 (for the Macintosh only) was released by Forethought, the small start-up company that developed it, as recounted by Parker (2001) in his history of the software. Originally named Presenter, the software generated black-and-white text and graphics pages that could be printed and converted into overhead transparencies via a photocopier. Shortly after the 1987 product launch, Microsoft acquired Forethought and by 1993, PowerPoint—now integrated with Word and Excel into Microsoft's Office Suite—was the dominant presentation software tool on the market. The growing availability of laptop computers in the mid 1990s further increased the interest in projecting PowerPoint directly from the computer, thus bypassing paper printouts and transparencies altogether. During this period, as Parker (2001) informs us, PowerPoint introduced the (in)famous AutoContent Wizard to help users create their visual aids. He notes that this feature, though originally named facetiously, influenced people's understanding and use of the PowerPoint tool significantly. Gradually these evolving understandings and uses, we argue, shaped how people enacted the genre itself.

By 2001, PowerPoint had captured 95% of the market in presentation graphics, and Microsoft estimated that at least 30 million PowerPoint presentations were made every day (Parker, 2001, p. 81). As we can see in texts such as Munter and Russell (2002) and Munter (2003), computer-generated slide shows had now become the dominant visual-aid medium in business presentations, although drawbacks of this medium were regularly described in texts. Indeed, in the business lexicon, "PowerPoint presentation" has come to refer to a presentation made using a PowerPoint slideshow projected from a computer. Although the PowerPoint software had been used to generate transparencies for over a decade, this usage was not typically encompassed by common understanding of the term.

At the turn of the twenty-first century, PowerPoint presentations had become so ubiquitous that their use had generated a backlash aimed at both the tool itself and how it was perceived to have influenced the business-presentation genre. Clear evidence of such a backlash can be found in the growing number of

Figure 2. *Dilbert* cartoon about PowerPoint.

Note: DILBERT: © Scott Adams/Dist. by United Feature Syndicate, inc.

cartoons, such as the Dilbert strip shown in Figure 2, depicting the (usually ineffective) use of PowerPoint. A widely circulated early example of the backlash was the so-called Gettysburg PowerPoint Presentation, which Peter Norvig (1999) created using the AutoContent Wizard and posted on the Web. It shows how the rich rhetoric of the Gettysburg Address would have been flattened and oversimplified by presenting it in standard PowerPoint format (see Figure 3). Other critical articles began appearing, such as Stewart's (2001) "Ban It Now! Friends Don't Let Friends Use PowerPoint," Schwartz's (2003) "The Level of Discourse Continues to Slide," and Tufte's (2003b) "PowerPoint is Evil: Power Corrupts. PowerPoint Corrupts Absolutely." This last article is the product of statistician and graphics guru Edward Tufte, perhaps the most vocal critic of PowerPoint presentations. He has self-published a 23-page essay entitled *The Cognitive Style of PowerPoint* (Tufte, 2003a), in which he argues that PowerPoint "slideware . . . reduces the analytical quality of presentations, . . . weaken[s] verbal and spatial reasoning, and almost always corrupt[s] statistical analysis" (p. 3). While demonstrating how strong the opposition to PowerPoint texts has grown as the software has become pervasive, Tufte's argument is less persuasive because it conflates the use of graphics in written documents, such as articles and newspapers, with the use of graphics as visual aids in oral presentations, failing to distinguish between fundamentally different genres (the article and the oral presentation) and the recurrent situations in which they are enacted. Moreover, as with all technologies, it is not the technology per se but how it is used in practice that determines outcomes and consequences. We thus turn now to an empirical examination of the PowerPoint-presentation genre as it is enacted in several organizations, showing how its use was shaped by historically evolving norms of business presentations and emergent technologies such as personal computers, laptops, telecommunications, and the PowerPoint software.

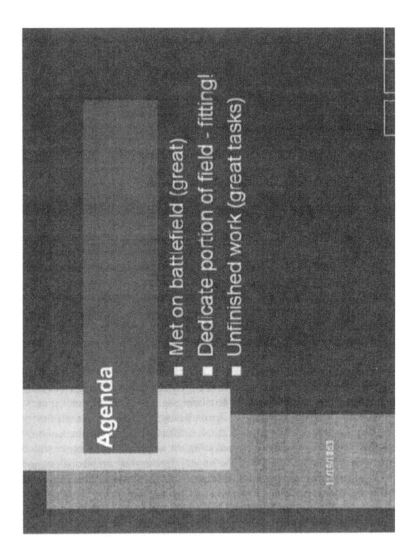

Figure 3. Agenda slide of the "Gettysburg PowerPoint Presentation" (Norvig, 1999).

THE GENRE OF POWERPOINT PRESENTATIONS

The PowerPoint business presentation as a genre is familiar to most people who have any contact with the business world. Like the report or the memo (also fairly broad genres), it has many specific variants that are used in marketing presentations, product-development presentations, progress reports, performance-results announcements, and so on. In spite of variations, however, it has a set of discursive characteristics that are, to use Schryer's (1993) terminology, "stabilized-for-now" and consequently recognizable to most business audiences. We will discuss these characteristics in terms of the six aspects of communication we developed above. As part of this discussion, we will also examine the emergence of some corollary genres being enacted with PowerPoint texts in many of the firms we have studied.[3]

Purpose

The PowerPoint presentation is typically used to inform, persuade, or motivate an internal or external organizational audience. Within firms, presentations may be used to propose a plan, explain a new program, or solicit input. For example, as described in Woerner, Orlikowski, and Yates (2005), we studied a regional facility management group (FacilityEast) within a firm we call Hardware Inc. Internal PowerPoint presentations were conducted by members of the headquarters facilities division presenting new programs and structures to FacilityEast and by members of FacilityEast presenting progress updates to the headquarters division. At several advertising firms we studied (see Kellogg, Orlikowski, & Yates, 2006), PowerPoint presentations of proposed advertising campaigns, aimed ultimately at the client, were often presented to internal groups as they were being developed, allowing those working on them to solicit comments from colleagues and to make changes before the client presentation. As further examples, Brooks (2004) discusses the use of PowerPoint presentations at a large systems-engineering firm, classifying them into three primary types by purpose and content: technical talks, intended to inform the audience about a new technology; get-acquainted talks, in which the speaker presented his or her group to another group in the hopes of generating collaboration; and project or program reviews, intended to communicate project/program status and to account for resources used.

Across firms, PowerPoint presentations are often used as ways of communicating with clients (potential or actual). Pitches—presentations proposing an approach to a potential client—are designed to secure business. During the

---

[3] This section draws in part on in-depth field studies we have conducted with Kate Kellogg, Heinrich Schwarz, and Stephanie Woerner, using techniques such as participant observation and interviews. It also draws on dissertation work by our former students Nils Fonstad and Natalia Levina.

dot-com bust, members of Adweb, an advertising firm specializing in Web-based marketing, were spending much of their time developing pitches for new business, as explained in Kellogg et al. (2006). Such pitches were critical to the survival of the firm. Even after the business had been contracted, members of the team working on a particular client project frequently presented their work-in-progress to the client in a PowerPoint *preez*, as they were called at Adweb. Such presentations both informed about and advocated their approach, while also soliciting feedback on the ideas as they were being developed.

Corollaries to the PowerPoint-presentation genre may have slightly or greatly different purposes. In one study, Fonstad (2003) described a common practice in consulting firms called "ghost sliding."

> The ghost sliding process involved creating emergent artifacts (e.g., rough sketches of slides) and iteratively building on, adapting, and receiving feedback on them until they become the final deliverables. Ghost sliding was a process that the consultants practiced extensively to develop presentations of findings for their clients. (p. 331)

This process was similar to the internal presentation of, and solicitation of comments on, material ultimately aimed at an external client or customer at the advertising firm, but it took the process a step farther. Here, the purpose was to help the consultant discover the findings and evidence needed in the final presentation.

We have also found that in many consulting firms, the written report that traditionally served as a final deliverable to the client (sometimes in conjunction with an oral presentation) has been replaced with a PowerPoint deck—that is, the stack of paper printouts of PowerPoint slides. Rather than the more formal, stand-up presentation with PowerPoint projection characteristic of the PowerPoint-presentation genre, the consultant now sits at a table with the client (typically represented by a small group of individuals) and walks them through the deck orally. At the end of this walk-through, the consultant leaves the deck, rather than a written report, as the primary project deliverable. This informal presentation practice and the PowerPoint deck challenge aspects of both the PowerPoint-presentation and the business-report genres. In particular, the deck of PowerPoint slides is expected to serve two different purposes: first, to function as a visual aid supporting an oral (informal) presentation; and second, to perform as a stand-alone deliverable (in many cases the only deliverable) reporting the results and conclusions of a project. PowerPoint texts created with this dual purpose generally have too much content to be effective presentation aids (which should support rather than overshadow the speaker), according to the genre norms of the business presentation, and too little content and context, including too few references and appendixes, to fulfill expectations for the report genre.

Use of this corollary genre generates considerable ambiguity around the appropriate form to use in order to accomplish the new purpose. The deck is usually easier to create but less detailed and nuanced than a report. It is easier for the audience to absorb during the presentation, but lacks the contextual details helpful for later use or for use by audiences not present at the original presentation. One senior manager at Adweb noted that the brevity constraints of PowerPoint texts offer some advantages over reports.

> I've always been appalled at the number of large documents that we'll give clients in an instance where PowerPoint would be a better delivery mechanism backed up by the large document. . . . You can't hand the senior executive a 200-page document.[4]

In this case, the manager saw value in PowerPoint texts serving as a sort of executive summary that complemented a more extensive accompanying report. Where the PowerPoint texts substitute for the report, the consequences are often less constructive. For example, in its conclusions, the board that investigated the Columbia shuttle accident criticized the pervasive use of PowerPoint slides within NASA (2003).

> During its investigation, the board was surprised to receive similar presentation slides from NASA officials in place of technical reports. The board views the endemic use of PowerPoint briefing slides instead of technical papers as an illustration of the problematic methods of technical communication at NASA. (p. 191)

The emerging multiple purposes of PowerPoint texts reveal the artful ways in which people are able to adapt and improvise their use of organizing structures. But even as the resultant corollary genres enable new communicative practices, we also see that the decoupling of texts from their original purposes and contexts is, at least initially, unsettling, generating tensions and ambiguities in expectations and use.

Content

Even though the specific content of PowerPoint presentations varies by purpose and situation, the genre entails expectations of content type. Thus, a PowerPoint presentation on a particular topic is typically expected to include some (but not too many) details about that topic. The quantity of information delivered in an oral PowerPoint presentation is usually less than that of a written

---

[4] This and other quotations from Adweb are from interviews conducted as part of our study of how members of Adweb used technologies in their work practices. See Kellogg et al. (2006) for more details about research methodology.

report. This tendency toward concise content may be both enabling and constraining. For example, a senior Adweb manager saw PowerPoint presentations as focusing attention on the critical issues: "I think [PowerPoint] forces people to try to decide what they think is important and to dare to be wrong." Others report finding the content of PowerPoint presentations to be too abbreviated, resulting in loss of meaning; as a technical manager at Adweb noted,

> I don't usually use Powerpoint. [The client liaison team] may ask us to send them an e-mail with a few bullet points for a Powerpoint preez. They don't want anything fancy, just a few points. . . . But sometimes things get so reduced to bullet points that both [they] and the client don't understand what we really mean.

Within contexts where PowerPoint decks are generated as primary deliverables from some activity and then passed on to people unfamiliar with the activity, the content may be overwhelming. In her study of a Web-based systems development firm, Levina (2001, p. 195) reports that members joining an existing project team were "brought up to speed" in a "knowledge dump" meeting where they were walked through a paper deck of the final PowerPoint deliverable from the planning phase of the project. Members described their experiences of this meeting as "death by PowerPoint," and subsequently referred to the large PowerPoint deck as the "Slide Graveyard" (p. 208).

Another set of content expectations has to do with the structure of the presentation, which has, on the most basic level, an introduction, a body, and a conclusion. This basic structure is common in many genres (including the business presentation before PowerPoint), but is now inscribed in the software's AutoContent Wizard. The typical introduction includes a slide previewing the talk's structure and sometimes its content, which is embedded in the AutoContent Wizard's generic presentation as a slide entitled "Topics of Discussion," appearing right after the title and introduction slides. In our empirical studies, we find that agenda slides abound but are not universal. For example, at Adweb, presentations always had an agenda slide to preview their structure; in Hardware, Inc.'s presentations, however, such agendas were not always present. The presentation's body can follow any of a number of structures depending on purpose and context, several of which are available in the AutoContent Wizard. Adweb pitches for creating interactive Web sites always covered the proposal's creative concept, technology requirements, proposed timeline, and deliverables. PowerPoint slide shows are usually projected in a fixed order, making it more difficult for the speaker to easily rearrange the slides during the presentation. This fact marks an important distinction between the PowerPoint presentation and presentations made using overhead transparencies. PowerPoint critic Tufte (2003b) refers to this as the tool's "relentless sequentiality, one damn slide after another" (p. 118). Finally, the conclusion of a generic presentation as embedded in the AutoContent Wizard is a slide focusing on Next Steps, and we found a final side focusing on the future common in many of the presentations we observed. This structure enables easier creation and comprehension of PowerPoint presentations, but the strong sequentiality also constrains the presenter's ability to respond flexibly to the local audience's interests and issues.

The firms we studied followed a common practice of reusing content from previous presentations in creating new PowerPoint texts, rather than beginning from scratch. At Adweb, the developers of new presentations typically began by copying content from earlier presentations. This practice enabled more efficient creation of new presentations while also facilitating some continuity with prior concepts and approaches. It also had constraining consequences, in particular creating difficulties for the graphic designers (known as Creatives) in the firm who were responsible for aesthetic design. As one Creative member commented,

> I've been trying to convert the Powerpoint preez that Joe [a project member who is not a Creative] has been using into a form that we can work with. I want the original artwork because if you cut and paste from someone else's preez you inherit all of their inconsistencies and you waste a lot of time.

As we saw within many contexts, once the PowerPoint content had been created, it often acquired a second life after (or in place of) its use as a visual aid to support an oral presentation. In this derivative reuse of content, we see the dual influences of the past, simultaneously facilitating efficiency and continuity in addition to restricting creativity and innovation.

Form

Expectations of form in PowerPoint presentations center on both the presentation as a whole and the PowerPoint text itself. The standard form of such presentations involves a single person standing before a group of people and using the PowerPoint slideshow to project visual aids onto a screen. At Adweb, a common presentation was the so-called pitch preez, in which members of the firm presented their proposal for work to prospective clients in the manner just described. In practice, however, presentations were not always delivered in this mode. In our studies, we often found that the presenter sat at a table with a small group of people and walked them through a deck, composed of paper copies of the slides. In some cases, decks were simply distributed to individuals without even a walk-through or discussion. One advertising firm we studied depended almost exclusively on such decks, in-house and with clients, relying on them as one of the firm's primary project deliverables.

Other variations in form included sending the PowerPoint file electronically to other sites and talking through the slides over an audio or video channel (e.g., telephone or video conference) as all parties viewed the slides. In this case, the audience was co-present with the presenter temporally but not spatially, and the presenter typically had limited control over the pace at which the audience saw the slides. When a collaborative tool such as NetMeeting was used, the speaker could control the display of the PowerPoint slides more tightly. Such a practice was common in Hardware, Inc., where each location participating in a distributed meeting could see the slides, either on a computer screen or projected onto a large screen, only as the presenter metered out the slides, one by one. Another common variation was placing a PowerPoint file on a Web site for people to view at different times. In this final example of a corollary genre, the slides themselves have to carry more of the substance of the presentation and thus need considerably more content than they would have if they were intended for projection by a speaker who would orally provide additional details and nuance about content and context.

Many expectations of the form of PowerPoint presentations center on the visual aids themselves. The PowerPoint software tool enables presenters to create certain types of visual effects but constrains them in their ability to go beyond what is offered by the tool. The tool offers an array of templates, allowing presenters to pick from a variety of background patterns, fonts, and colors. The templates offered are quite elaborate, but many are too intricate and distracting to fit the guidelines of contemporary experts in this area. While colors within templates can be changed, such change takes time and is somewhat difficult to do. Furthermore, creating unique templates is more difficult than using those provided by the software, and presenters tend to be discouraged from doing so very often. At Hardware, Inc., a small set of standard slide templates had been developed for use in most internal presentations. In general, internal presentations in most of the firms we saw were less likely to use elaborate and unusual formats than external presentations that were designed to catch the eye of a customer or client.

Beyond the templates provided by the PowerPoint software, other form expectations are also embedded in the tool. For example, when the creator of a PowerPoint file opens a new slide, he or she is given multiple options from which to choose, including a title slide, a bulleted list (the most common form used for most of the slides we saw), a graph, a table, and a blank slide into which clip art, photographs, or other images could be inserted or drawn. The tool both draws on previous norms (e.g., bullet points rather than complete sentences) and institutionalizes them by providing them as simple-to-use, built-in features. A variety of transitional devices (sounds as well as visual effects) are offered for creating slide shows. Builds, in which bulleted list items appear one at a time, are also a much-used feature of PowerPoint slide shows, adding to the "relentless sequentiality" Tufte (2003b) sees in such presentations (p. 118). Similarly, the auto shapes and clip art allow easy creation of slides with images as well as text, a feature that both enables and constrains. Using some visual elements rather than all text fits prescriptions that predate PowerPoint and can make the presentation more interesting. Moreover, builds, if used strategically, can avoid the dilemma that made earlier presenters slide paper down overhead transparencies, revealing the points a line at a time. However, all of these features are often used mindlessly and inappropriately, irritating and distracting the audience. Tufte (2003a, 2003b) notes, among other complaints, that the three-dimensional effects often used in graphs distort relationships, while the sounds and transitions quickly become annoying to those who see many such presentations. For designers at Adweb, where such PowerPoint presentations were used for almost all communication with the client, and as the basis for internal collaboration, the Creatives disliked the constraining aspects of the tool, as evident in these quotes from two members:

> I'm not a big fan of PowerPoint. I hate it that we use such a primitive program. It [is] such a low-end presentation mechanism for a Creative deliverable. . . . It doesn't differentiate us.

> You know, the problem with these programs is that they are built for novices. And for people who know how to do design, they don't let you do what you want to. It's really frustrating. . . . Now we never get to use the handcrafting [custom design of fonts and letter spacing] that I learned in school. Now everything is done with programs, and it takes the art and true design out of it.

For these design professionals, the constraining qualities of PowerPoint were more salient than the enabling ones. They found it difficult to exercise the skills and techniques that they believed gave them a distinctive identity and afforded value to the firm. Not surprisingly, they both resisted and resented the use of PowerPoint on Adweb projects. In spite of this opposition, however, project managers continued to use PowerPoint extensively. Because they wished to achieve a consistent look and feel to both internal and client presentations, they also insisted that everyone in the firm use PowerPoint and its built-in features.

Participants

In its most typical form, the PowerPoint presentation is assumed to be given by a single individual, usually standing at the front of a room, and directed toward a physically present audience of more than two or three people. Thus the roles in the typical presentation are clear: one speaker controlling the pace and visual aids and several co-present listeners who may, depending on cultural norms, ask questions of and interact with the presenter. Brooks (2004) analyzes the enactment of PowerPoint presentations as a rite in the systems-engineering firm she studied and sees the audience as participating in "synchronized collective action " directed toward a common symbolic focus of attention: the projected PowerPoint slide show itself (p. 70).

When corollaries to the PowerPoint-presentation genre are enacted, these roles may shift somewhat. One shift occurs when the audience is no longer in the same location. For example, members of the face-to-face audience may have more interactions with the speaker than those at a distant site. When two locations are sharing a screen through NetMeeting (as often occurred at Hardware, Inc.), control can, with the permission of the person with the slides, be transferred back and forth. Thus the presenter may sometimes give over control to the audience, blurring the roles of speaker and listener. Individuals viewing a set of PowerPoint texts in slide show mode, but without a live presenter, control the pace of the presentation themselves, therefore playing a more active role than the more typical audience, although they are constrained by not being able to obtain clarifications and elaborations while viewing the slides.

The creation of PowerPoint presentations may also involve different people, as anyone with access to the PowerPoint tool may create slides, unlike in the past when transparencies were generally created by intermediaries such as designers or secretaries. While a presenter may create the presentation alone, such sole authorship was rare in our studies. Collaborative authorship was more common, with two different modes being evident. First, in the distributed mode, multiple internal parties contributed different sections of the presentations. At Adweb and other advertising companies we studied, the authorial control was distributed across multiple people, and the PowerPoint file (stored centrally on the firm's network) served as the vehicle for collaboration in preparing the presentation. Sometimes, as in the consulting company studied by Fonstad (2003), clients were involved in the process to produce the collaborative presentation.

> Ghost Sliding was an iterative process where quick, rough representations of each slide were drawn up and discussed with the client to develop consensus on what statements were going to be included in the presentation and what data needed to be collected to support those statements. After each discussion, more data were collected and more detail and specificity added to the slides. Then, after a significant number of changes had been made to the presentation-in-progress, the next version was discussed with the clients. (pp. 224-225)

Second, the centralized mode of presentation preparation tended to follow hierarchical lines, with the presentations being created by junior people, secretaries, or even a specialized design staff (though the popularity of PowerPoint has made such organizational arrangements less common) and then worked over, extended, and edited by the more senior individuals. The consulting company in Fonstad's (2003) study used this mode of creating PowerPoint texts: an office assistant would receive (via fax) handwritten drafts of slides from consultants working at client sites, translate these into presentation visuals on her computer, and then return the PowerPoint texts to the consultants as an e-mail attachment.

Time and Place

We will look at time and place together, since these expectations are typically tightly coupled. The standard PowerPoint presentation occurs within a single room where the speaker and the entire audience are assembled at the same time, achieving temporal and spatial alignment or what Zerubavel (1981) calls symmetry. Brooks (2004) describes the extreme symmetry she observed in PowerPoint presentations at her site as synchronized collective action. The presenter talks and the audience listens in real time, simultaneously watching the slides as they are projected on a screen. Using PowerPoint projection also allows the presenter to make changes in the slides up to or even during the presentation, enabling the use of more up-to-date visuals.

Although the standard PowerPoint presentation occurs in one time and one place, a key feature of the PowerPoint tool is the ease with which electronic files of the PowerPoint texts can be sent to other locations or saved for other times. These texts are often shipped to another location so a distant audience can view them on their computers or another projection screen with the speaker's voice-over. In several companies we have studied, presentations using PowerPoint slides were delivered over telephone conferences, so that the target audience (e.g., the client for Adweb or the local office representative for a regional meeting in Hardware, Inc.) saw the slides and heard the voice but never saw the speaker. In conjunction with telephone conferences, participants often accessed PowerPoint texts through collaborative tools such as NetMeeting or WebEx, which enabled people at different locations to share a screen controlled by the presenter. However, when these systems failed to work properly (e.g., due to software, hardware, or telecommunications errors), the constraining aspects of distributed presentations became very apparent. We saw this happen a few times at Hardware, Inc. As a member of the FacilityEast group pointed out, "I mean, as great as NetMeeting is, occasionally, it's our worst enemy because [if] you rely on it so much and it's not working for you, it pretty much almost shuts down the entire meeting."

In some of the corollary genres we observed in our research there were shifts in both temporal and spatial expectations away from the PowerPoint-presentation genre. For example, PowerPoint texts were often printed out as decks, one slide to a page, or in the handout form suggested by the tool (with 2–6 slides per page), and provided as hard copies to be taken away after the talk (as well as looked at during it). As noted previously, these decks or so-called take-aways often replaced reports as deliverables and could be viewed at a different time and place by the audience members or by their colleagues when the context, details, and nuance provided orally by the speaker were absent. When the creator of the PowerPoint texts could no longer count on being present to interpret and amplify them, he or she often put many more words and images on each slide than could realistically be absorbed by the viewer watching the presentation in real time. So creating the same PowerPoint texts for use as part of both the PowerPoint presentation as well as the corollary deck-as-deliverable genre produced the dual problems of information overload and loss of meaning referred to before.

Another corollary genre we saw frequently at Hardware Inc.—the online slideshow available on the Web and viewed by one person at a time—is typically even more distributed temporally and spatially, making it more difficult for viewers to achieve a common understanding. While all the viewers could see the same PowerPoint slides, their temporal and spatial separation precluded a collective dialogue about the slides' meanings and implications. Even when the slides contained more text than was functional for a real-time presentation, they frequently did not have enough content and context to address the inevitable ambiguities and reservations that arose, which led to different understandings by different viewers and, because of the lack of synchronicity, allowed no opportunity for discussion that might have resolved the confusion.

IMPLICATIONS FOR WORKPLACE COMMUNICATION

In this chapter, we have drawn on genre theory and our empirical research to suggest that the PowerPoint presentation has emerged as a powerful and complex communicative structure that both reflects and shapes organizational practices while also enabling and constraining a range of social actions and outcomes. By focusing on how its multiple discursive expectations are enacted in use, and how these change with the emergence of corollary genres within and across multiple media, we can begin to understand this genre's dynamic influences on and consequences for organizational life.

Tufte (2003a, 2003b), Norvig (1999), and others have noted some consequences of the constraints that the PowerPoint tool imposes on presenters, including the limited, fragmented, and flattened content appearing in bulleted form. Indeed, we see consequences for the audience (and sometimes even for the presenter) that include limited comprehension, information overload ("death by PowerPoint"), lack of reflection, idea fragmentation, and reductionism. At the same time, our empirical studies also demonstrate that the tool enables as well as constrains. In particular, we saw that it facilitates distributed coauthoring of content, as well as collaborative development of ideas iteratively over time. It also encourages discursive focus and brevity, thus forcing "people to try to decide what they think is important and to dare to be wrong," as the senior Adweb manager quoted earlier noted. Similarly, the strong linearity of most PowerPoint presentations is shaped by the sequentiality of slides and the difficulty of viewing them in any other order. Although this sequentiality reduces the speaker's responsiveness to the audience, it promotes a strong narrative line, aiding the "repurposing" of the PowerPoint texts into decks-as-deliverables and stand-alone slide shows on the Web.

The repurposing of the PowerPoint texts in corollary genres has its own set of consequences for communication in organizations. Using a particular component of the presentation genre—the PowerPoint texts—in a variety of recurrent situations, many of which have very different discursive requirements and social purposes, poses many communicative challenges. Printing out PowerPoint texts as decks or handouts poses difficulties for readability because text that shows up well in projection mode sometimes is impossible to read in black-and-white print, and because it eliminates the role of transitions and special effects. While the limitations of printed PowerPoint texts constrain some uses of the PowerPoint tool, they also help curb some of the worst excesses of the PowerPoint slide shows as enacted with computer projection. The bigger problems occur, however, when the PowerPoint texts are repurposed for use in a setting where temporal and spatial symmetry are no longer present (the deck-as-deliverable, the Web-based presentation, etc.). Because they anticipate secondary uses without knowing their specifics, those who create decks typically crowd them with more content than is necessary or effective for a visual aid accompanying a live presenter, creating information overload for the live audience. Nevertheless, such stand-alone presentations, as they are still called (even though no presenter accompanies the visual aids), lack the more detailed context and nuanced content of a live presentation or of a written report. Thus they contribute to communicative ambiguity and loss of meaning. This problem is likely to increase as the PowerPoint texts are used at a greater temporal or spatial remove from the original presentation. In a commentary for *Slate Magazine*, Kaplan (2003) quotes Edward Mark, a historian for the U.S. Air Force, as observing

> Almost all Air Force documents today, for example, are presented as PowerPoint briefings. They are almost never printed and rarely stored. When they are saved, they are often unaccompanied by any text. As a result, in many cases, the briefings are incomprehensible.

As electronic or paper renditions of PowerPoint texts become the only record of major activities, comprehension is reduced and organizational memory deteriorates. Levina's (2001) Slide Graveyard is at the same time both too much information and too little for new organization members. Just as audience expectations from visual aids have been shaped by the use of multiple expressive media available in the PowerPoint texts—text, charts, images, animation, audio, and video—expectations may also change as PowerPoint texts are transmitted and viewed in different transmission media: computer-projection, overhead transparency, paper copy, and electronic file. Moreover, audience expectations for the live presenter change as that presenter is co-present or is mediated by videoconference or telephone. When the presenter is absent altogether, as seems to be increasingly the case in corollary genres such as the online slideshow for individual viewing or the take-away deck-as-deliverable, the genre expectations become increasingly uncertain. Until clearer expectations arise around these corollary genres, we can expect continued genre ambiguity, communicative difficulty, and discursive experimentation.

CONCLUSION

We have suggested here that corollary genres emerge from microlevel improvisations that shift some of the genre expectations associated with a particular genre, but do not (yet) transform it. Such shifts, to borrow from Pollan (2002), "inflect the prose of everyday life without rewriting it" (p. 142). Seen as inflections in conventional discursive practice, the concept of corollary genres helps us to articulate the process through which knowledgeable human agents begin to modify and experiment with aspects of their established communicative genres. As derivatives of established genres, corollary genres begin to decouple texts from the particular recurrent situation around which they emerged, thus enabling as well as constraining new forms of discursive expression. In enacting such shifts of conventional discursive practice, human agents produce a variety of tensions and ambiguities, challenging their communicative effectiveness. But in doing so, they also generate possibilities for social change.

ACKNOWLEDGMENTS

We appreciate the helpful comments of Mark Zachry and Charlotte Thralls on an earlier draft of this chapter. This research was supported in part by a grant from the National Science Foundation (award #ITR-0085725).

REFERENCES

Barley, S. R. (1986). Technology as an occasion for structuring: Evidence from observation of CT scanners and the social order of radiology departments. *Administrative Science Quarterly, 31,* 78-108.

Brinton, W. C. (1914). *Graphic methods for presenting facts.* New York: Engineering Magazine Co.

Brooks, J. (2004). *Presentations as rites: Co-presence and visible images for organizing memory collectively.* Unpublished doctoral dissertation, University of Michigan, Ann Arbor.

Connelly, W. J. (1958). Mechanical aids to the effective presentation of technical papers. In T. E. R. Singer (Ed.), *Information and communication practice in industry* (pp. 166-194). New York: Rheinhold.

DeKay, S. H. (2003). *The historical evolution of the employment resume in the United States, 1950-1999.* Unpublished doctoral dissertation, Fordham University, New York.

Donnet, N. (1988). *Power presentations on the business stage.* Toronto: Gage.

Fonstad, N. O. (2003). *The roles of technology in improvising.* Unpublished doctoral dissertation, Massachusetts Institute of Technology, Cambridge.

Giddens, A. (1993). *New rules of sociological method* (2nd ed.). London: Hutchinson/ Cambridge: Polity.

Giddens, A. (1984). *The constitution of society: Outline of the theory of structure.* Berkeley: University of California Press.

Holcombe, M. W., & Stein, J. K. (1990). *Presentations for decision makers.* New York: Van Nostrand Reinhold.

Kaplan, F. (2003, June 4). The end of history: How e-mail is wrecking our National Archive. *Slate Magazine.* Available at http://slate.msn.com/id/2083920/.

Kellogg, K., Orlikowski, W. J., & Yates, J. (2006). Life in the trading zone: Structuring coordination across boundaries in post-bureaucratic organizations. *Organization Science, 17*(1), 22-44.

Krell. E. (2001). Finance breaks out. *Business Finance.* Retrieved April 27, 2004, from http://www.businessfinancemag.com/magazine/archives/article.html?articleID=13696.

Lemke, J. L. (1995). *Textual politics: Discourse and social dynamics.* Bristol, PA: Taylor & Francis.

Lessing, L. P. (1950). The story of the greatest chemical aggregation in the world: DuPont. *Fortune 42,* 86-118.

Levina, N. (2001). *Multi-party information systems development: The challenge of cross-boundary collaboration.* Unpublished doctoral dissertation, Massachusetts Institute of Technology, Cambridge.

Miller, C. R. (1984). Genre as social action. *Quarterly Journal of Speech, 70,* 151-167.

Morris, J. O. (1972). *Make yourself clear! Morris on business communication.* New York: McGraw-Hill.

Munter, M. (1982). *Guide to managerial communication.* Englewood Cliffs, NJ: Prentice-Hall.

Munter, M. (1992). *Guide to managerial communication* (3rd ed.). Englewood Cliffs, NJ: Prentice-Hall.

Munter, M. (2003). *Guide to managerial communication* (6th ed.). Upper Saddle River, NJ: Prentice-Hall.

Munter, M., & Russell, L. (2002). *Guide to presentations.* Upper Saddle River, NJ: Prentice-Hall.

NASA. (2003, October). *Columbia accident investigation board: Final report.* Retrieved April 28, 2004, from http://caib.nasa.gov.

Norvig, P. (1999). The Gettysburg PowerPoint presentation. Retrieved April 28, 2004, from http://www.norvig.com/Gettysburg/index.htm.

Orlikowski, W. J. (2000). Using technology and constituting structures: A practice lens for studying technology in organizations. *Organization Science, 11*(4), 404-428.

Orlikowski, W. J., & Yates, J. (1994). Genre repertoire: Examining the structuring of communicative practices in organizations. *Administrative Science Quarterly, 39,* 541-574.

Parker, I. (2001, May 28). Absolute PowerPoint. *The New Yorker,* 76-87.

Piper, H. A. (1938). Manuscript of address made to the National Office Management Association, New York Chapter. Hagley Museum and Archives (accession 1662/#73), Wilmington, Delaware.

Pollan, M. (2002). *The botany of desire.* New York: Random House.

Robbins, L. M. (1985). *The business of writing and speaking.* New York: McGraw-Hill.

Schein, E. (1969). *Process consultation.* Reading, MA: Addison Wesley.

Schryer, C. F. (1993). Records as genres. *Written Communication, 10,* 200-234.

Schryer, C. F. (2002). Genre and power: A chronotopic analysis. In R. Coe, L. Lingard, & T. Teslenko (Eds.), *The rhetoric and ideology of genre: Strategies for stability and change* (pp. 73-102). Cresskill, NJ: Hampton Press.

Schwartz, J. (2003, September 28). The level of discourse continues to slide. *New York Times, 4,* 12.

Stewart, T. (2001, February 5). Ban it now! Friends don't let friends use PowerPoint. *Fortune,* 210.

Tufte, E. (2003a). *The cognitive style of PowerPoint.* Cheshire, CT: Graphics Press.

Tufte, E. (2003b, September 9). PowerPoint is evil: Power corrupts. PowerPoint corrupts absolutely. *Wired Magazine,* 11.09.

White, J. (1996). *Color for impact: How color can get your message across—Or get in the way.* Berkeley, CA: Strathmoor.

Woerner, S., Orlikowski, W. J., & Yates, J. (2005). Scaffolding conversations: Combining media in organizational communication. *Proceedings of the European Group for Organizational Studies Conference,* Berlin.

Yates, J. (1985). Graphs as a managerial tool: A case study of DuPont's use of graphs in the early twentieth century. *Journal of Business Communication, 22,* 5-33.

Yates, J. (1989). *Control through communication: The rise of system in American management.* Baltimore, MD: Johns Hopkins University Press.

Yates, J., & Orlikowski, W. J. (1992). Genres of organizational communication: A structurational approach to studying communication and media. *Academy of Management Review, 17,* 299-326.

Yates, J., & Orlikowski, W. J. (2002). Genre systems: Structuring interaction through communicative norms. *Journal of Business Communication, 39,* 13-35.

Yates, J., Orlikowski, W. J., & Okamura, K. (1999). Explicit and implicit structuring of genres: Electronic communication in a Japanese R&D organization. *Organization Science, 10,* 83-103.

Yoshioka, T., Yates, J., & Orlikowski, W. J. (2002). Community-based interpretive schemes: Exploring the use of cyber meetings within a global organization. *Proceedings of the Hawaii International Conference on System Sciences, Hawaii, 35,* pp. 1-10.

Zerubavel, E. (1981). *Hidden rhythms: Schedules and calendars in social life.* Chicago, IL: University of Chicago Press.

CHAPTER 5

Reason and Rationalization: Modes of Argumentation Among Health-Care Professionals

Martin Ruef

Building on the social-action framework of Max Weber, this chapter considers how the structure of professional work engenders different cultural claims of rationality in public discourse. I begin by identifying three dimensions of rationality—scientific, instrumental, and formal—and show how they may be manifested in modes of argumentation among professionals. My theoretical propositions suggest that the level of abstraction and objectivity (i.e., orientation toward descriptive rather than prescriptive claims) in professional discourse increases with the structural distance of a professional group from the domain of practice. I apply the social-action model to a systematic database of discourse among health-care professionals between 1975 and 1994, with particular emphasis on texts that discuss aspects of formal organization in the health services domain. My evidence suggests that there are structural hierarchies that regulate claims of abstraction and objectivity among physicians, administrators, policymakers, nurses, and allied health professionals. I conclude the chapter by discussing possible relationships between the culture of these professions and exercises of inter-professional power.

One of the principal communicative skills that professionals have at their disposal is their mode of argumentation, that is, their logic for presenting knowledge, value, and action claims considered independently of the content of those claims. Even a cursory review of professional discourse reveals that verbal or written statements can advance similar propositions but be structured in radically different ways, employing disparate modes of argumentation or methodology. An editorial in a professional nursing journal may tout the cost-effectiveness of health maintenance organizations (HMOs) on the basis of anecdotal evidence offered by practitioners, while a policy analysis article may promote the same message, offering statistical evidence or a more theoretical motivation. A research monograph read by health economists may present a purely descriptive comparison of costs in HMOs versus other settings, whereas a text read by executives may prescribe a procedure whereby efficiencies can be realized through HMO contracts.

At a deeper level, these varying modes of professional argumentation tend to be grounded in different assumptions about rationality and how social action ought to be regulated by it. In this respect, a useful (and foundational) account of rationalization was offered by the sociologist Max Weber (1968; original publication date 1924), who considered how authority—the legitimate ability to regulate communicative and other social action—is tied to different conceptions of rationality. Weber posited that a series of premodern forms of authority had culminated in "legal rational" authority, which relied "on a belief in the legality of enacted rules and the right of those elevated to authority under such rules to issue commands" (p. 215). Weber associated this development of legal authority with a process of rationalization in which action is increasingly regulated by abstract, means/ends calculation. Subsequently, a number of sociologists have noted the importance of professional culture in producing rationalization in contemporary society (Abbott, 1988; DiMaggio & Powell, 1983; Parsons, 1954, pp. 36-37).

In this chapter, I will sketch a perspective on professional communication that underscores the relationship of rationality and authority. The principal questions to be answered are twofold: (1) how are claims of communicative rationality linked to the social structure of professionals; and (2) are certain forms of rationality more successful in establishing professional jurisdictions than others? Using health-services professionals (construed broadly to include physicians, nurses, administrators, policymakers, etc.) as an empirical exemplar, I will offer analyses to explore the first question and provide a theoretical motivation for the second. I argue that both issues are linked through a hierarchy of abstraction and "objectivity," whereby the relationship of professionals to the domain of practice regulates the character of their communication and, consequently, the strength of their jurisdictional claims.

With respect to empirical context, I choose to focus on the U.S. health-care sector between 1975 and 1994, a tumultuous period of contestation in professional jurisdictions, both in terms of control over treatment and control over the organization of service delivery (see Begun & Lippincott, 1993). The chapter places particular emphasis on the rationalization processes and jurisdictional claims associated with organizational knowledge, thus casting a wider net on the task domain than is entailed by medical treatment or diagnosis *per se*. A focus on organizational knowledge also dovetails with the classic Weberian

(1968) interest in bureaucratic rationalism and the systems of written rules that it entails (pp. 957-958).

FACES OF RATIONALITY

Within the social sciences, the concept of rationality has often appeared in multiple guises, leading to a lack of clear criteria for operationalization and empirical inquiry. In one relevant passage, Weber (1946; original publication date 1915) points out that rationality

> means one thing if we think of the kind of rationalization the systematic thinker performs on the image of the world: an increasing theoretical mastery of reality by means of increasingly precise abstract concepts. Rationality means another thing if we think of the methodical attainment of a definitely given and practical end by way of an increasingly precise calculation of the adequate means. (p. 293)

In this discussion, Weber was not simply identifying two forms of rationality, but three (see Table 1). First, as Kalberg (1980) notes, he calls attention to theoretical abstraction as a manifestation of rationality (pp. 1152-1155). Weber's "systematic thinker" (for instance, a health-policy researcher) seeks to determine highly generalizable principles whereby features of reality can be described or predicted. In his research on professional jurisdictions, Abbott (1988) has referred to this as a form of contentless abstraction that allows professionals to generalize across instances and across time. This manifestation of rationality can be denoted under the rubric of *scientific rationality*. Second, Weber (1946) calls attention to purposive, means/ends decision making as another manifestation of rationality. As Scott (2001) notes, whether viewed narrowly, in terms of maximizing an objective function, or more broadly, in terms of "satisficing" behavior exhibited by boundedly rational agents, this conception has probably had the strongest impact on contemporary social science (pp. 66-68). I will use the term *instrumental rationality* to describe this alternative manifestation.

Table 1. Dimensions of Rationality and Modes of Argumentation in Professional Discourse

Dimension of Rationality	Modes of Argumentation
Scientific	**Abstract** versus **Concrete**
Instrumental	**Instrumental** versus **Descriptive** versus **Substantive**
Formal	**Formal** versus **Informal**

Weber's quote also subtly introduces a third form of rationality that is not clearly differentiated from its scientific and instrumental counterparts. In the second sentence of the quotation, he mentions the use of "precise calculation," alluding to the implicit role of mathematization or formal logic in regulating mean/ends decision making. Such formal rationality—whether construed simply in mathematical terms or, more broadly, in terms of precise, replicable heuristics—is neither a necessary nor sufficient condition for instrumental rationality. Actors employing mean/ends rules may rely on an extensive stock of tacit knowledge, which is not immediately subject to precise quantification; observers attempting to emulate the instrumental rationality of others may not find reproducible formulae. And of course, abstract formalization in and of itself, according to Kalberg (1980), need not automatically entail scientifically rational procedures. Accordingly, I will treat formal rationality separately from scientific and instrumental rationality.

My preliminary overview thus yields three distinct dimensions of rationality: scientific, instrumental, and formal. To see how these dimensions relate to modes of argumentation among health-service professionals, it will be useful to consider some empirical examples. In the first example below, the discourse fragment exhibits all three forms of rationality. This fragment, extracted from B. E. Fries and A. S. Ginsberg (1979), invokes a formal, instrumental, and abstract mode of argumentation.

> The increased demand of unscheduled visits to hospital outpatient facilities, and particularly to walk-in clinics, necessitates the consideration of means to control this flow and to reduce in-hospital waiting times. A model is developed in which a percentage of the unscheduled visits are assumed to be delayable, e.g., patients with non-urgent complaints who call may be asked to delay their arrivals for specified lengths of time. A simple rule for determining, dynamically, the length of this delay was examined by computer simulation. The results demonstrate significant reduction of in-facility waiting times while only marginally increasing the time patients wait from their initial contact with the clinic until seen by a practitioner. . . . (pp. 967-972)

In this passage, the authors propose a formal specification of patient arrivals at outpatient facilities, develop quantifiable delay rules, and investigate their implications by using a computer simulation. They lay out an instrumental, means/ends schema to control the flow of unscheduled patient visits and reduce waiting times. Finally, they advance a (scientifically) abstract model that is not limited to a particular hospital or outpatient facility but that is, in principle, generalized for any such facility. The hospital manager who chooses to adopt the suggested patient admission protocol is to some extent compelled to adopt the calculative, instrumental, and universalizing discourse of the authors.

The discourse fragment can be contrasted with another from the *New England Journal of Medicine* (Berarducci, Delbanco, & Rabkin, 1975), which exhibits an informal, descriptive, and concrete mode of argumentation:

> Hospitals have become major providers of primary medical care by default rather than design because of the decreasing availability of primary physicians in community practice. In 1972, Beth Israel Hospital in Boston closed its traditional general clinics and established a primary-care group practice—Beth Israel Ambulatory Care. The practice is staffed by teams of full-time salaried physicians, other health professionals and paraprofessionals, house staff and medical students. More than 25% of patient visits are managed by nonphysicians . . . (pp. 615-620)

As Habermas (1984) would describe this fragment, the authors are engaged in a descriptive mode of argumentation rather than an instrumental mode; the text describes a situated state of affairs and not necessarily a set of instructions that serve as a recipe for implementing any particular procedure (p. 329). The text is also informal, relying on case study methodology rather than mathematical analysis; and concrete, emphasizing the particular instance of Beth Israel Hospital as opposed to an empirical generalization of all American hospitals. In these respects, the text can be said to deviate from rationality, as viewed along the three dimensions of instrumentality, formality, and scientific abstraction.

These two texts do not circumscribe the full range of professional modes of argumentation. Weber (1968, pp. 85-86) also points to "substantive rationality" as an orientation toward social action that differs from both instrumentality and description. Indeed, many debates among professionals are prescriptive in nature, but they are not limited to the myopic means/ends reasoning inherent in instrumental action. The discourse fragment from *Medical Care* (Lewis, 1976) serves as an example of an informal, substantive, and abstract mode of argumentation.

> An effective national health insurance program should provide adequate, continuous, and comprehensive coverage for all. In framing current proposals, it seems that policy makers have not adequately considered differences in medical care utilization and needs between men and women. The inequities which result occur primarily because women now have two central sources of medical care—an obstetrician/gynecologist and a general practitioner or internist—while men have only one. The article delineates four issues of particular importance to women: 1) eligibility provisions which insure women through their husbands' policies; 2) benefit structures which exclude aspects of reproductive health services and/or fail to explicitly recognize women's two central sources of care; 3) provider certification provisions which exclude free-standing clinics and/or nonphysician personnel; and 4) incentives for reform of health delivery which force women to choose between their two current sources of care. (pp. 549-558)

In this case, the author employs a substantive mode of argumentation, establishing criteria of success for a national health-insurance program and, in particular, calling attention to the issues that may impede quality or equality in health-services delivery for women. Rather than simply dealing with a means/ends (instrumental) form of rationality, the text concentrates on ends and the obligatory dimension of social life, saying, for example, that "an effective . . . program *should* provide adequate, continuous, and comprehensive coverage for all" (emphasis added). With respect to other aspects of rationality, the fragment displays mixed modes of argumentation—on the one hand, it is informal in orientation, avoiding mathematical analysis; on the other hand, it is abstract and universal.

Modes of argumentation in professional discourse can thus be arrayed along three dimensions of rationality, each dimension falling into one of seven modal categories enumerated in Table 1. I refer to professional discourse as being formalized to the extent that it relies ostensibly on replicable, logical, or quantitative heuristics. For present purposes, this sense of formal rationality includes the use, or advocated use, of techniques from the fields of accounting, legal methodology, biometrics, operations research, statistics, pure mathematics, computer science, and the like. Professional discourse is said to be abstract to the extent that it is not concerned with the idiosyncrasies of individual cases, but seeks to develop universals (generalizing over instances and over time). For present purposes, such scientific rationality refers to an attempt to generalize the conclusions of a text to the entire U.S. health-care sector. Finally, professional discourse can be said to be instrumental to the extent that it advances instructions for particular procedures or techniques to be adopted by the audience of the discourse. Within the boundaries of such instrumental rationality, one can include all texts that are prescriptive and that limit their purview to a consideration of specific means rather than broader substantive ends.[1]

THE STRUCTURE AND DYNAMICS OF RATIONALIZATION

I now turn to the structural bases and dynamics of rationalization processes, paying particular attention to the institutional peculiarities of the health-services domain, while also drawing out more general theoretical propositions. In many respects, the changes in the health sector over the past few decades can be seen as a quintessential example of a rationalization process. In the immediate post-World War II era, the decisions guiding health-services administration, as Burns

---

[1] The typology of rationality offered here maps closely onto that developed earlier by Kalberg (1980). He identifies four dimensions—formal, theoretical, practical, and substantive—with the first two corresponding to my concepts of formal and scientific rationality, respectively, and the latter two representing the nondescriptive modes within my instrumental dimension.

(1990) points out, were rooted in a community-oriented logic that shunned abstract conceptions of organizing care. Following the more recent rise of managed-care arrangements, instrumentalism, formalism, and scientific management seem to have usurped the modes of argumentation in professional discourse (Scott, Ruef, Mendel, & Caronna, 2000). As Weberian scholars have recognized, however, the march of rationalization does not tend to proceed in a uniform, linear fashion (Käsler, 1988, p. 172). The regulation of workplace communication is often on the front lines of contestation between professional factions, and its pendulum swings reflect this tenuous position.

Professional Classes and Rationality

My general conception of the social structure of the health-care field is based on a tripartite division of professional labor composed of (1) technical professionals (including physicians, nurses, and allied health workers in the medical field); (2) managerial professionals (health-facility administrators, business consultants, management analysts); and (3) institutional professionals (social scientists, policy analysts, legal consultants). Technical professions are involved in the transformation of production inputs (e.g., sick patients) into outputs (healthy patients). Managerial professions are charged with obtaining and administering resources for the production system. Institutional professions relate the system to its broader environment, including the external regulatory structures that monitor its activity as well as the general public (Parsons, 1960).

The three professional classes are broad and obviously admit to more refined differentiations. In particular, the spectrum of professionals and paraprofessionals in the technical cluster is substantial, ranging from radiology technicians and nurses to doctors of medicine (MDs) and osteopathy (DOs) (see Scott & Lammers, 1985). In order to capture some of these distinctions, the technical professionals can be split into two groups for heuristic purposes, with an elite technical subclass subsuming physicians (both generalists and specialists) as well as clinical researchers, and a nonelite subclass subsuming nurses, dentists, technologists, and other allied health professionals.

The Hierarchy of Abstraction

Given the tripartite distinction of professional classes, one might ask: how is rationality regulated by the social structure of health-care professionals? Let us begin with abstraction (both scientific and formal), which has been a traditional focal point in perspectives on professional power. A perspicacious explanation is offered by conceptualizing a hierarchy of abstraction among the three professional classes: technical, managerial, and institutional. Generally speaking, one may expect the level of abstraction regulating the discourse of a professional class to be proportional to the structural distance of that class from the tasks and phenomena described by the discourse. For instance, nurses, physicians, and allied health workers, who must deal with organizational aspects of health delivery on a very concrete basis, are likely to display a low level of abstraction in their discourse on the topic. Their high level of direct involvement with patients and other technical workers encourages this orientation. By contrast, social scientists and policymakers, who enjoy considerably more structural distance from the day-to-day organization of health delivery, are likely to display a high level of abstraction. Their structural distance is most acutely exemplified in the lack of patient contact among institutional professionals; as Weber noted, the severance of personal bonds is often a prerequisite for rational abstraction. Finally, managerial types will tend to display an intermediate level of abstraction as a result of their moderate distance from concrete processes of health-care delivery.

The mechanisms whereby the hierarchy of abstraction comes into being are based on the situated character of professional practice. The physician, nurse, or social worker who witnesses the organization of patient care in its full complexity may be loathe, for instance, to assign a simple number to the level of satisfaction enjoyed by patients; such a reduction of practice to a mathematical formalism seems to strip away a wealth of qualitative, contextual knowledge. By the same token, generalization of satisfaction statistics over a population of patients may be seen as inimical to the local knowledge of technical professionals. Policy experts, health economists, and other institutional professionals who lack such local knowledge are more likely to dismiss it in favor of convenient abstractions, whether formal or theoretical.

Hierarchy of Objectivity

Considering the remaining dimension of rationality (instrumentalism), the structural explanation becomes slightly more complex. This case involves three modes of argumentation—descriptive, substantive, and instrumental—rather than two. The distinction between the purely epistemic stance provided by the descriptive mode (identifying an "objective" state of affairs) and the interventionist stance provided by the instrumental and substantive modes (oriented toward changing means and changing ends, respectively) calls conceptual attention to a hierarchy of objectivity that parallels the hierarchy of abstraction introduced above. The level of "objectivity" (tendency toward descriptive argumentation) regulating the discourse of a professional class is expected to be proportional to the structural distance of that class from the tasks and phenomena described by the discourse. Thus we anticipate that institutional professionals, who have the largest structural distance from the domain of practice (e.g., the least patient contact), will be most likely to engage in purely descriptive discourse, while technical professionals, who are intimately connected to the domain of practice, will be most likely to engage in prescriptive (instrumental or substantive) discourse. Managers tend to display a mixed mode of argumentation in this respect.

The hierarchy of objectivity points to something of a paradox when seen through the lens of classical theories of rationality. Structural factors tend to promote both instrumental and substantive claims in the social groups at the bottom of the hierarchy. By contrast, as Kalberg (1980) points out, Weber stressed the potentially conflicting nature of instrumental and substantive rationality (pp. 1164-1165). Nevertheless, professional discourse provides many exemplars that show those individuals who are most familiar with the instrumental means of production are also most concerned with the ends to which those means will be devoted; for example, the nuclear physicist who warns against the proliferation of weapons of mass destruction or the molecular biologist who worries about the long-term implications of genetic research. Insofar as technical professionals are not alienated from their means of production (i.e., not standing in a purely instrumental relation to those means), we can expect that they will also claim a good deal of substantive rationality.[2]

Temporal Dynamics of Rationalization

When turning from the equilibrium aspects of rationality, which are rooted in professional social structure, to the temporal process of rationalization, the historical context of professional work takes center stage. Two aspects of this context will be of primary concern in this section: (1) what is the technological mode of production within which professional work is situated; and (2) what is the level of interdependency (both functional and geographic) of professional work within an organizational field?

Dynamics of Abstraction

Health-care technology has experienced a fairly dramatic transition over the past few decades, shifting from a primary emphasis on therapeutic modalities (treatment of patients after the fact) to an emphasis on diagnostic/monitoring modalities (recognition of risk factors before the fact). Evidence for this trend can be found in the structure of Medicaid payments. According to the *Health Care Financing Review* (1980–1990), increases in expenditures for diagnostic laboratory services outpaced increases in payments for prescription drugs by a ratio of almost two-to-one during the 1980s. These developments in patient care have been mirrored by investments in monitoring tools for health-services administration. In the 1960s, electronic data sets and computational technology in the field had barely passed their stage of infancy (see Fuchs, 1996 on the early history of health-services research). By the 1990s, tools for formal abstraction

[2] In general, one would expect alienation to be far more pronounced among nonprofessionals than among professionals or paraprofessionals. Accordingly, it is unclear whether the hierarchy of objectivity can be applied to account for communicative practices in other, less professionalized sectors.

had become widely disseminated. Government agencies now expend considerable resources in collecting data and analyzing the performance of providers and insurers. New specialty areas, such as medical informatics, have sprung up to support the creation and dissemination of information resources and computationally based diagnostic technologies. These areas have not only proven hospitable to workers from nonmedical backgrounds, such as computer science and applied physics, but have also given rise to new types of professionals such as biomedical engineers. According to Barley (1990), technologists that were once limited to the sidelines of the health-care field (or subservient positions under the watchful gaze of physicians) are now exercising more control as the sector moves toward formalization.

All of this would seem to suggest a secular upward trend in the formal abstraction of health-services discourse. More theoretically, this empirical generalization relates to an important question concerning the correspondence between social-structural and cultural change. Aside from certain anomalous situations, according to Eriksson (1975), a correspondence principle holds that structural and cultural variations tend to parallel one another. Given increases in the information resources and professional labor force (computer scientists, medical engineers, statisticians, etc.) that are devoted to systematic data collection and analysis within the health-services domain, the correspondence principle predicts an increase in the formalization of professional discourse in the field.

The temporal dynamics of scientific rationality in health care can be hypothesized to follow a similar upward trend. There is considerable evidence of universalizing tendencies in the culture of the field. The idiosyncratic particularities of professional work in health care that were identified by researchers in the 1960s, including problems of moral hazard and negative externalities, are increasingly dismissed in more recent approaches to studying the sector (Brown, 1986). By the same token, local knowledge governing the administration of facilities, in particular communities, is increasingly abandoned in favor of more general principles of health-services organization. In terms of far-flung linkages among health-care facilities and integrated delivery systems, such cultural changes parallel the extent to which the sector itself has evidenced both greater external structural interdependence with other market sectors and more internal interdependence. These developments suggest further support for the correspondence principle, noted above, in the form of a link between the universalism of professional discourse and the structural interdependency (internal and external) of health services.

EMPIRICAL APPLICATION

Data

The textual data for this chapter were extracted from MEDLINE®, the largest and most systematic database of machine-readable text in the medical arena. Over a thousand professional journals and conference proceedings published between

1975 and 1994 were scanned with a search engine for content related to the domain of health-services organization and policy.[3] The journals subsumed major publications targeted at physicians and allied health professionals (e.g., *New England Journal of Medicine, Journal of the American Medical Association,* etc.), those targeted at facility managers and business consultants (*Modern Healthcare, Health Care Management Review*), and those oriented toward policy professionals and social scientists (*Milbank Memorial Fund Quarterly, Journal of Health Politics, Policy, and Law*). A subset was extracted based on the following criteria: (1) texts must include full abstracts; (2) they must appear in English-language journals; and (3) they must either be published in the United States or appear in international journals that offer coverage of the American health-care sector. The resulting database features 5,971 texts.

Using the Science Citation Index (SCI), the Social Science Citation Index (SSCI), and qualitative analyses of journal mission statements, all texts in the database were classified by professional readership into one of sixteen categories (see Table 2). These were aggregated further into four larger classes of professionals, including so-called elite technical professionals, nonelite technical professionals, managerial professionals, and institutional professionals, as shown in the table. Ruef (1999) offers detailed information on coding and analytical methodology.

Results

Based on a systematic analysis of the MEDLINE® abstracts, Ruef (1999) developed a quantitative profile of the modes of argumentation found in texts oriented toward different professional audiences. The orthogonal projection in Figure 1 summarizes my findings concerning the structure of argumentation among health-care professionals. Circles indicate point estimates for structural parameters, while dotted boxes delineate intersections of confidence intervals around the estimates.[4] With respect to formalization, one immediately notes that institutional professionals are clearly differentiated from other types of professionals in terms of the high level of quantitative discourse that they are typically exposed to. Along the same dimension, elite technical and managerial types achieve more of an intermediate level of formalization, a requirement for strong jurisdictional claims. However, these groups are only weakly differentiated from nonelite technical professionals. As a consequence, we observe what might be referred to as a "partial" hierarchy of abstraction along this dimension.

---

[3] The search engine used includes both the Aries MEDLINE Knowledge Finder and customized programs developed by the author.

[4] Confidence intervals suggest that 95% of discourse for a given professional audience falls within those boundaries. The omitted parameters (constrained to zero) are those corresponding to elite technical professionals.

Table 2. Classification of Texts in the Corpus (1975–1994)

Journal classification	No. of texts (%)	Journal type
Business and General Administration	748 (12.5)	Managerial
Clinical and Basic Research	96 (1.6)	Technical*
Dentistry	82 (1.4)	Technical
Drug and Alcohol Addiction	29 (0.5)	Technical
Educational Administration	240 (4.0)	Managerial
Law	142 (2.4)	Institutional
Medicine, General	968 (16.2)	Technical*
Medicine, Specialized	1114 (18.7)	Technical*
Mental Health	414 (6.9)	Technical
Nursing and Social Work	354 (5.9)	Technical
Pharmacology	160 (2.7)	Technical
Public Health	483 (8.1)	Institutional
Public Policy	856 (14.3)	Institutional
Rehabilitation	49 (0.8)	Technical
Social Science	163 (2.7)	Institutional
Technology	73 (1.2)	Technical

*"Elite" technical professionals.

We may also note here a partial hierarchy of abstraction for scientific rationality. Compared to elite technical professionals, institutional and managerial types are more likely to engage in abstract modes of reasoning, constituting claims that are scientifically rational in this respect. By contrast, nonelite technical professionals are slightly less likely to advance abstract claims than their elite counterparts, instead maintaining concrete modes of argumentation. But the managerial and institutional task domains are not differentiated from one another along this dimension; indeed, abstraction is slightly higher among the texts directed at managers.

As Figure 1 shows, structural analysis of professional communication reveals a fairly clear hierarchy of objectivity. With respect to the substantive argumentation, the omitted reference category (including physicians and clinical researchers) has the highest parameter coefficient, with non-elite technical and managerial professionals offering somewhat fewer substantive claims and institutional professionals offering significantly fewer substantive claims. We may

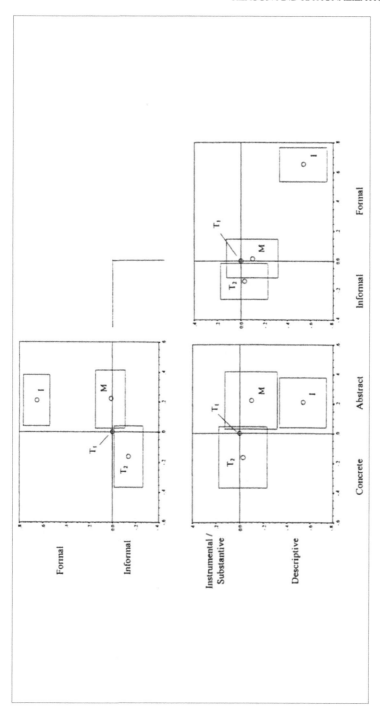

Figure 1. Structure of argumentation in health-services discourse.

Legend: I (Institutional Professions); M (Managerial Professions); T₁ (Technical Professionals [Physicians]); T₂ (Technical Professions [Others])

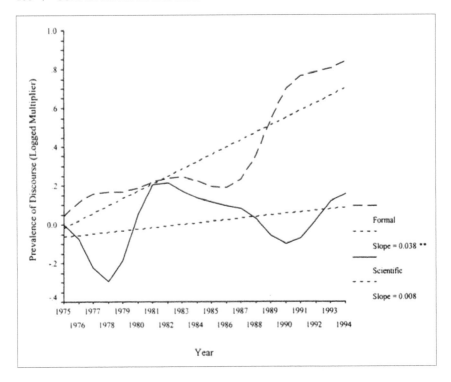

Figure 2. Dynamics of formal and scientific rationality in
health-services discourse.

**Positive time trend is statistically significant at *p* < .01 level.

identify a very similar pattern for instrumental argumentation, except that non-elite technical professionals rank higher than elite technical professionals in their tendency toward means/ends reasoning.[5]

Figure 1 necessarily omits the dynamics of rationalization over the period from 1975 to 1994. Annual fluctuations in formal rationality, as Figure 2 indicates, support the proposition that the level of formalization would experience a secular increase over the period, paralleling structural developments in terms of the number of professionals active in the clinical information systems area, as well as resource investments in medical informatics. This tendency in the health-services discourse provides some evidence for a correspondence principle between structural and cultural evolution in this field.

---

[5] As intimated earlier, the propensity toward more instrumentality (and lower substantive rationality) among non-elite technical professionals may result from greater alienation than their elite counterparts.

Scientific rationality also increases slightly over time, but large fluctuations render the effect statistically insignificant. The historical inflection points are of some interest, though, since they match major paradigm shifts in theories of health-services organization and policy. A relative decline in scientific abstraction between 1975 and 1978, for instance, corresponds broadly to decreasing faith in the national health-insurance proposals of the 1970s, while the following increase in abstraction maps onto the development of market-oriented discourse for organizing health-care delivery. Within that cycle, abstraction reaches its peak shortly before the passage of the Prospective Payment System for Medicare reimbursement in 1983;[6] thereafter, the discourse becomes more concrete as the details of the new market paradigm are elaborated. In the early 1990s, scientific abstraction again increases as the Clinton Health Security Act (1993) presents new challenges to the existing market ideology. Thus, rather than exhibiting a monotonic rise with structural integration in the health-care field, scientific abstraction more closely maps onto a dialectic, in which abstract rules are developed with paradigm emergence, while more concrete principles arise during paradigm development and decline.

DISCUSSION

In this chapter, I have sketched a basic account of rationalization processes among health-care professionals. Quantitative analyses of modes of argumentation in the health-services discourse have been employed to consider the structure and dynamics of formal, scientific, and instrumental rationality. The findings suggest support for a partial hierarchy of abstraction that regulates discourse among technical, managerial, and institutional professionals and stronger support for a "hierarchy of objectivity." I also found evidence for some correspondence principles that apply between the evolution of rationality and structure in the health-care field, in particular insofar as formal rationality is concerned.

Before considering the implications of these findings for the exercise of inter-professional power, it will be useful to review some of the general sociological arguments connecting dimensions of rationality with jurisdictional strength. Following Wilensky (1964, pp. 148-149), Abbott draws an inverse-U-shaped relation between scientific rationality and professional power, noting that professional jurisdictions are weakened by too much abstraction or too much concreteness. Abbott (1988) states that optimal abstraction involves a balance where "contrary forces . . . push abstraction in professional knowledge toward an equilibrium between extreme abstraction and extreme concreteness" (p. 104).

---

[6] The Prospective Payment System (PPS) shifted Medicare from retrospective reimbursement of costs to prospective payments linked to diagnosis. It is widely credited with the decentralization of decision-making processes in the sector (ProPAC, 1989).

The positioning of professionals relative to this equilibrium dictates whether jurisdictional claims will be accepted or rejected. Changes in the equilibrium can be an important social dynamic regulating authority among a system of professions.

Although Abbott restricts his claims to scientific rationality, they may be extended to formal and instrumental rationality as well. Formal rationality involves a trade-off between concreteness and abstraction, though it takes on a slightly different character than that suggested by the preceding discussion. Consider the hospital manager who seeks to present an evaluation of competing health-care providers in her market niche. She could rely heavily on a formal market analysis, in which data on competitors are gathered, statistical models are executed, and competitive metrics are communicated. This presentation may require relatively little concrete knowledge, but it will require a great deal of time devoted to data gathering and analysis. At the other extreme, the hospital manager could rely exclusively on a large stock of informal, qualitative rules that she has acquired as a result of her extensive experience in a health-care market, while avoiding numerical analyses. Discourse systems that involve a balance of formal and informal strategies tend to maximize the jurisdictional strength of the communicators, due to the diminishing returns that accrue to each strategy individually.

Considered as a whole, the structure of professional argumentation (see Figure 1) offers theoretical insights into the strength of professional jurisdictions and the potential power exercised by the different professional classes. Managers, in particular, have been well-positioned to influence organizational decision making in the sector, mediating between technical and institutional professionals along one dimension of rationality (instrumentality), siding with technical professionals on another (formal rationality), and matching their argumentation to that of institutional professionals on a third dimension (scientific rationality). Meanwhile, the technical professionals have consistently been down in the proverbial trenches, while the institutional professionals have been up in the clouds. According to Scott et al. (2000), when the professions are considered in this light, it is not surprising that earlier paradigms of health-services organization, such as professional dominance and federal intervention, were defeated by a market-oriented paradigm in the 1980s. In particular, once institutional professionals had introduced certain elements of formal and scientific rationality, these efforts at rationalization were turned against them, adopted by managerial professionals, and reformulated in more concrete and prescriptive ways.

The initial accomplishments of institutional professionals during the late 1960s and the 1970s are explained to a considerable extent by the dynamics of rationalization. Early on, the lack of adequate information resources and sectoral integration allowed institutional professionals to gain an advisory jurisdiction in the health-care arena, bringing with them tools for formalization based in scientific rationality. As the growth of formal and universal principles increased in the late 1970s and early 1980s, however, the abstract knowledge claims pushed by policymakers, legal consultants, and social scientists began to seem excessive. The overall increase in formal and scientific rationality had opened a new niche in the discourse of the sector, calling for communication that mediated between extreme concreteness and extreme abstraction. This niche could then be exploited by entrepreneurial managers.

CONCLUSION

A long-standing tradition in the sociology of the professions has treated capacity for communicative rationality as a central feature of professional life (Abbott, 1988; Wilensky, 1964). The formal and theoretical rationality of professional discourse distinguishes physicians, lawyers, and academics from car mechanics and waitresses. Rationality is the defining feature that regulates professional authority, allowing professionals to maintain monopoly over certain task domains or niches. As Abbott (1988) puts it, "abstraction is the quality that sets interprofessional competition apart from competition among occupations in general . . . [a]bstraction enables survival in the competitive system of professions" (pp. 8-9; see also Wilensky, 1964). Although generally congenial to perspectives on interprofessional power stressing market competition (e.g., Begun & Lippincott, 1993), this cultural framework calls attention to the importance of discourse systems in creating jurisdictions as opposed to the material resources emerging from professional associations, licensure, training schools, and university programs. The framework's emphasis on publicly mediated communication also serves to differentiate it from approaches that stress interpersonal networks among professionals or paraprofessionals (e.g., Barley, 1990).

While promising new theoretical insights, the cultural perspective has seen limited empirical implementation, largely due to a failure of social scientists to attend to the properties of professional communication. As is often the case, this methodological problem may be symptomatic of deeper theoretical issues. Professional communication is not an isolated source of data, but is linked fundamentally to rationality claims being advanced in modern society. This suggests that the analysis of professional jurisdictions and power can profit from being embedded in a broader institutional sociology of rationality processes, as pursued by Weber (1946; 1968), Habermas (1984), and neoinstitutional scholars (e.g., DiMaggio & Powell, 1983).

From this perspective, future research should address several questions in greater detail. At a microlevel, how do hierarchies of rationality in professional discourse regulate professional interaction, collegiality, conflict, and domination? What sources of contextual variation (e.g., workplace environment) serve to accentuate or decrease the effect of different claims of rationality? At a macrolevel, how do different professional groups leverage claims of rationality to lobby for preferred organizational forms, codes of conduct, or legislation? To what extent is the strategic efficacy of their discourse regulated by the ecology of communicative claims being advanced by other interest groups? Issues such as this provide fertile ground for further research on the implications of communication for professional authority and influence.

REFERENCES

Abbott, A. (1988). *The system of the professions: An essay on the division of expert labor.* Chicago, IL: University of Chicago Press.

Barley, S. (1990). The alignment of technology and structure through roles and networks. *Administrative Science Quarterly, 35*, 61-103.

Begun, J., & Lippincott, R. (1993). *Strategic adaptation in the health professions.* San Francisco, CA: Jossey-Bass.

Berarducci, A. A., Delbanco, T. L., & Rabkin, M. T. (1975). The teaching hospital and primary care: Closing down the clinics. *New England Journal of Medicine, 292*, 615-620.

Brown, L. (1986). Introduction to a decade of transition. *Journal of Health Politics, Policy and Law, 11*, 569-583.

Burns, L. (1990). The transformation of the American hospital: From community institution toward business enterprise. *Comparative Social Research, 12*, 77-112.

Clinton, W. (1993). *Health Security Act.* Washington, DC: Government Printing Office.

DiMaggio, P., & Powell, W. W. (1983). The iron cage revisited: Institutional isomorphism and collective rationality in organizational fields. *American Sociological Review, 48*, 147-160.

Eriksson, B. (1975). *Problems of an empirical sociology of knowledge.* Stockholm: University of Uppsala.

Fries, B. E., & Ginsberg, A. S. (1979). The effect of delay rules in controlling unscheduled visits to hospitals. *Medical Care, 17*, 967-972.

Fuchs, V. (1996). Economics, values, and health care reform. *American Economic Review, 86*, 1-24.

Habermas, J. (1984). *The theory of communicative action* (2nd ed., Vols. 1-2). Boston, MA: Beacon Press.

Health Care Financing Review (HCFA). (1980-1990). *Health care financing review.* Baltimore, MD: Health Care Finance Administration.

Kalberg, S. (1980). Weber's types of rationality: Cornerstones for the analysis of rationalization processes in history. *American Journal of Sociology, 85*, 1145-1179.

Käsler, D. (1988). *Max Weber: An introduction to his life and work.* Chicago, IL: University of Chicago Press.

Lewis, D. A. (1976). Women and national health insurance: Issues and solutions. *Medical Care, 14*, 549-558.

Parsons, T. (1954). *Essays in sociological theory.* Glencoe, IL: Free Press.

Parsons, T. (1960). *Structure and process in modern societies.* Glencoe, IL: Free Press.

Prospective Payment Assessment Commission (ProPAC). (1989, June). *Medicare Prospective Payment System and the American Health System: Report to Congress* (Book XII). Washington, DC: The Commission.

Ruef, M. (1999). *The rise of managed health care: An inquiry into the evolution of discourse, ideology, and power.* Unpublished doctoral dissertation, Stanford University, Stanford.

Scott, W. R. (2001). *Institutions and organizations* (2nd ed.). Thousand Oaks, CA: Sage.

Scott, W. R., & Lammers, J. (1985). Trends in occupations and organizations in the medical care and mental health sectors. *Medical Care Review, 42,* 37-76.

Scott, W. R., Ruef, M., Mendel, P., & Caronna, C. (2000). *Institutional change and health care organizations: From professional dominance to managed care.* Chicago, IL: University of Chicago Press.

Weber, M. (1946). The social psychology of the world religions. In H. Gerth & C. W. Mills (Eds.), *From Max Weber: Essays in sociology.* New York: Oxford University Press. (Original work published 1915.)

Weber, M.. (1968). *Economy and society: An interpretive sociology.* In G. Roth & C. Wittich (Eds.), New York: Bedminster. (Original work published 1924.)

Wilensky, H. (1964). The professionalization of everyone? *American Journal of Sociology, 70,* 137-158.

CHAPTER 6

Writing and Relationship in Academic Culture

Kenneth J. Gergen

Within academic culture the written text is not only the major means of communicating what we take to be knowledge, but a significant form of relating. In my view, the regulatory practices of writing within academic culture often contribute to alienated, narcissistic, and hierarchical forms of relationship. Moreover, they insulate the disciplines from each other and the academic world from cultural life more generally. In this light it is useful to consider emerging genres of expression and their potentials for transforming relationships. Attention is directed, in particular, to transparent and relational forms of expression. In the case of transparent writing, the text opens the writer to the audience. The embodied and multiplicitous character of the writer is made available. In relational writing, uses of dialogue and performance invite the audience to join the writer in exploration. Both genres challenge existing regulations and invite new and more congenial forms of academic life.

To be able to dance with one's feet, with concepts, with words: need I still add that one must be able to do it with the pen too?
Friedrich Nietzsche, *The Gay Science*

Primary words do not signify things but intimate relations.
Martin Buber, *I and Thou*

In large measure we differentiate among cultural groups in terms of recurring patterns of relationship. Patterns of intimate relationships, gender, family life, friendship, community ties, and social hierarchies, for example, are all used as significant cultural markers. Because of their centrality to social life, communicative practices are pivotal to our understanding of any culture. My concern in this offering is with the communicative practices that constitute the academic or scholarly culture. The character of discursive practices within academic culture has been the focus of intense concern in the social sciences. This concern has largely grown from social constructionist scholarship and can be traced in its most recent history to the sweeping ramifications of Thomas Kuhn's 1962 volume, *The Structure of Scientific Revolutions*. Although reflecting earlier lines of thinking within the sociology of knowledge and history of science, Kuhn's writing most clearly articulated the importance of communicative practices in generating knowledge claims and associated values and methods of study. In effect, as scientific subcultures are formed through communication, so are paradigms of understanding generated. This abiding concern with the social construction of knowledge has stimulated a vast range of inquiry into the rhetorical, ideological, and interpersonal processes characterizing knowledge-generating cultures.[1]

The vast share of this inquiry into the creation of knowledge cultures has focused on the effects of communicative practices on outcomes, which is to say, the authoritative truth claims within the sciences. Far less attention has been given to the impact of communicative practices on forms of relationship within these enclaves. Within recent years such concerns have been reflected in communication studies, emphasizing in particular the ways in which conversation practices establish hierarchies of power and privilege (Krippendorf, 1993; Tracy, 1996). This chapter shifts concern from conversation to writing practices. By and large, a scholar's professional trajectory is determined by the quantity and quality of his or her writing. One's professional identity is primarily dependent on contributions to the literature. However, I wish to press beyond matters of acceptance and status to consider ways in which forms of writing shape personal relations within these cultures. As Linda Brodkey (1987) proposes, academic writing is primarily a social practice. In my view the forms of writing employed within scholarly culture instantiate key conceptions of the person, and in so doing, contribute to the character of relationships within this culture.

My primary focus will be on social-science writing. Writing, like speaking, is fundamentally an action within a relationship; it is within relationship that writing gains its meaning and significance, and our manner of writing simultaneously invites certain forms of relationship while discouraging or suppressing others. As I shall attempt to demonstrate, our traditional forms of writing are deeply problematic in their regulatory impact on scholarly culture and its societal potentials more generally.

---

[1] For an extended account of the development of social-constructionist thought in the past 30 years, see Gergen (1994).

GENRES OF WRITING/ FORMS OF LIFE

Once sensitized to writing as constitutive of relationship, we are prompted to ask what forms of relationship are invited by existing traditions of scholarly writing? How do such forms of writing regulate relations among colleagues, between teachers and students, and between participants in disparate scholarly communities? To appreciate what is at stake, consider John Shotter's (1997) commentary on contemporary academic exchange.

> Is there a kind of violence at work in intellectual debates and discussions; in the university colloquium, seminar, or classroom; in academic texts? Is there something implicit in our very ways of us relating ourselves to each other in academic life in present times that makes us fear each other? Is there something in our current circumstances that makes us (or at least some of us) anxious about owning certain of our own words, or taking a stand? Speaking from my own experience, I think there is. (pp. 17-18)

If forms of writing do contribute to the kinds of social worlds we inhabit, how does this process occur? In the present case I will be particularly concerned with the ontological and valuational presumptions reflected in various genres of writing. As I shall propose, our styles of inscription not only carry with them conceptions of the person, but as well images of ideal character—that to which we may properly aspire. Where topics of academic concern may change radically across time, forms of inscription often remain stable. Thus, for example, while scientific psychology has recently undergone a major shift from behavioral to cognitive models of human functioning, the forms of scientific writing have remained relatively obdurate. Most importantly for the present, in establishing a conception of the ideal person, modes of inscription lay the grounds for our forms of relationship. Thus, in our forms of writing, we contribute to the character of relations that make up scholarly culture. And indeed, by virtue of its broad educational mandate, academic culture exerts an influence on the culture at large. Forms of writing are wedded, then, to forms of cultural life (see also Gergen, 1994, chap. 7).

In my view, our major traditions of writing within the social sciences are deeply problematic in terms of their outcomes both for academic culture and the broader society which it serves. At the same time, we have experienced in the past three decades a slow but profound undermining of traditional assumptions in the social sciences. Indeed, this process of conceptual transformation is largely responsible for the cultural turn that anchors the essays in this volume. Thus, as the chapter unfolds I will treat emerging and iconoclastic forms of writing, both as they reflect changing conceptions of the person and their potential for transforming scholarly culture. As the reader will find, the unfolding account will also be accompanied by shifts in genres of exposition. The hope, then, is that the various forms of writing employed here will be self-exemplifying.

Before proceeding, two caveats are required. First, my focus on writing traditions within the social sciences is limited. While I do believe there are significant implications for both the natural sciences and humanities, there are particular circumstances within these latter realms that prohibit easy generalization. Second, while I will be critical of our major traditions of social-science writing, this is not to argue for their wholesale abandonment. My purpose here is to champion an expansion not a diminution of our potentials for representation and resulting relationship.

ENCAPSULATED SELVES: PRIVILEGE
AND PEJORATION

In significant degree the social sciences are progeny of Enlightenment discourse, derived from presumptions about the nature of human knowledge, cosmological order, and the potentials for human betterment through systematic inquiry. Most significant for present purposes, important elements of the Enlightenment conception of human functioning make their way into the forms of writing now prevailing in the sciences. With respect to images of human functioning, I am speaking here of the dualist tradition in which we presume the existence of individual minds capable of acquiring knowledge about an external material world. The central ingredient of the mental world, from Descartes to contemporary cognitive psychology, is the capacity for rational thought (now often referred to as "information processing"). In particular, when reason is linked with the sensory capacities of observation, the individual can accumulate objective knowledge of the world. Objectivity is impaired to the extent that desire, motivation or emotion (all expressions of one's animal instincts) alter the processes of reason and observation. It is on the basis of individual knowledge that the common person can rise above the animal kingdom, disclaim the authority of kings and popes, and survive—if not thrive—in the material world.

This view of individual functioning also carries with it significant implications for our forms of scholarly writing. Of particular prominence are the desired criteria of verbal economy, logical coherence, clarity, dispassionate demeanor, comprehensiveness, and certainty. The demand for efficient thought conduces to Occam's insistence on "no unnecessary words." Because logical minds contain no inconsistencies, coherence of argument is essential. Inasmuch as the knowledgeable mind is discerning, clarity of exposition is required. With desires and values suspect, a plain, flat, nonemotive style is prized. With vision of cumulative knowledgeable in place, then comprehensiveness of one's account is at a premium, and certainty (or the reduction of the realm of the obscure) is esteemed. It is not the "well-wrought urn" that serves as the guiding metaphor of social-science writing, but something more akin to the perfectly appointed gunboat—powerful in resources, flawless in operation, insistent on purpose, and beyond defeat by anyone. Consider the three exemplary specimens that follow.

1. If P's only want in C at t is to achieve G, and if P believes that trying to do A in C at t is the alternative with the highest likelihood of leading to G, if P believes that he or she can perform A in C at t, and if the alternative acts that P believes he or she can perform are believed by P not to involve less exertion than A, then P will try to do A in C at t (Smedslund, 1988, p. 74).[2]

2. It is also predicted that when silence (in communication) occurs, it will be differentially assigned, on the basis of the rules as either (I) a gap before subsequent application of Rules 1(b) or 1(c), or (ii) a lapse on the non-application of Rules 1(a) (b) and (c), or (iii) a selected next speaker's significant (or attributable) silence after the application of Rule 1(a) (Levinson, 1989, p. 48).

3. People who were induced to recall their own behaviors relevant to a personality trait were not faster than people who merely defined the word to answer a subsequent question about whether they had the trait. Had recalling specific behavioral instances been part of the self-assessment process, the second question would have been answered faster after autobiographical memories than after a semantic task (Sia, Lord, Blessum, Thomas, & Lepper, 1999, p. 520).

At the outset, such writing reflects and sustains the presumption of the individual, bounded mind. The words find their origin within the psychological interior and serve as conduits for rendering the interior manifest. In this sense, such writing creates the sense of a social world that is fundamentally fractured; each of us exists as an isolated atom. The writing reflects the contents of one's own mind, distinct from the mind of those who have preceded (and thus the severe sanctions against plagiarism), along with those who may subsequently read. The writer is the Ursprung, the seer, the knower. Others are inherently alien.

In addition to the implicit division of society into autonomous entities, there is a structure of privilege built into the form of writing. Social-science writing represents itself as an advance in knowledge, that which has never yet been said, that which renders one's own discernment superior to the discernment of others. In contrast, the audience is positioned by the writing as ignorant or unaware. The rhetor never addresses an equally enlightened colleague. The form of address is that of revelation—of truth, reason, or inspired insight; a reader is thus required who "has yet to see." (My own writing here is no exception; the manner of my articulation creates me as a knowledgeable source, separated from you, the reader, now positioned as uninformed about these matters.) The hierarchy of privilege is also, by implication, an order of adequacy. When writing represents itself as knowledge, the writer is defined as adequate (rational, discerning,

---

[2] S. L. Levinson (1989, p. 48).

advanced) and the audience as less so. In effect, we inherit and sustain forms of writing that contribute to alienated relationships, to feelings of inadequacy, and to an atomistic and hierarchical conception of society.

These circumstances also lend themselves to bounded disciplines that are opaque and typically uninviting to those standing outside. In constructing a world of bounded and alienated being, the individual writer finds himself or herself in a condition of potential solipsism. There is no means of verifying the accuracy or rationality of his or her cerebration—no means of gauging its value as a contribution to knowledge. More broadly, the solitary individual lacks the capacity for self-authentication. Required, then, is a responsive audience, but more specifically an audience that adheres to the role of the ignorant. Largely owing to its state of ignorance, the audience will respond appreciatively. It is primarily through the existence of the appreciative audience, either real or imaginary, that the scholar is authenticated as an acceptable human being. Of course, by virtue of curricular demands one may secure a certain degree of approbation from one's students. Further, because colleagues understand the cultural rules of reciprocity, they may offer a certain degree of support: it is mainly by providing affirmation that one can secure it in return. Thus, by rationalizing a certain form of curriculum, and building a network of professional support, the isolated individual achieves a sense of value. More broadly, in order to sustain one's conception of self as a worthy being (within the Enlightenment mold), something approximating an academic discipline is required.

This centripetal lurch is aggravated by further factors. Self-authentication, as we have seen, typically (though not inevitably) requires an audience willing to accede to a role of subservience. Yet, for a scholar to fill the role of subservience is simultaneously to define himself or herself as an incomplete vessel who is unable to deliberate independently. Thus, the scholarly landscape is populated with those who are suspicious of or set against the writer, unless the latter is deceased or apostatized. For the mature scholar, critique is thus the principle form of rejoinder to one's colleagues. Inhering in our common forms of scholarly exchange, then, is essentially a vast undermining of confidence in individual worth. The scholar is confronted by an essential condition of self-uncertainty: "who am I, what is my value, how good am I?" To relieve the uncertainty, the cycle is repeated—new inquiry, new writing—but now in a more advanced form. New concepts may be constructed, unknown works brought to light, more obscure vocabulary extricated, new populations explored, all of which, in effect, increase the range of what there is to know. Such increments enhance one's position in the hierarchy, and in turn, send others into a spin of refutation and refurbishment. The conceptual, terminological, and methodological world expands rapidly; and to maintain the sense of individual value, there is little means of exiting the process. It is through continuous reading, critique, and reconstruction that the shaky grasp on worthy being is maintained. At the same time, an otherwise impenetrable wall of words is erected by the community.

Should the stranger struggle to enter this house of language, and employ the discourse with less facility, he or she risks derision. Contained within our mode of writing is thus our form of academic life and a telos of individual and disciplinary separation.

TRANSPARENT WRITING: OPENING TO THE OTHER

> How, then, do we write ourselves into our texts with intellectual and spiritual integrity? How do we nurture our own voices, our own, and at the same time lay claim to "knowing" something?
>
> Laurel Richardson (1997), *Fields of Play*

I am scarcely alone in my discontent with traditional writing practices along with the relationships they invite. Anxiety at professional meetings is especially acute. Few are brave enough to present their work in anything less than written form; to stumble or falter would reveal they might possibly be less than perfect. I have indeed known colleagues who memorize entire papers for a 20-minute, one-time presentation at a social-science conference. Of course, the steady drumming of spoken text soon has the audience either lost in fantasy or nodding off. Fortunately, the dialogues around the social construction of knowledge have begun to eat away at the assumptions that sustain this kind of communication. Constructionist arguments first remove the authority from those claiming superior knowledge: "They don't know more; they simply know different." Constructionist critique informs us that social-science writing is simply a preferred rhetoric within a particular community. There is no foundational reason to sustain a given tradition, and indeed, if our traditions of writing alienate us from each other we are free to pursue alternatives.

Many social scientists have largely abandoned the traditional forms of writing in search for alternatives of greater promise for a caring and open community. Most prominent in this regard are scholars attempting to mine the more familiar discourses of everyday life, especially those we recognize as revealing "the genuine person." Rather than positioning themselves as fully rational agents, bounded and superior, these scholars generate a more recognizably human persona in their writing, one to whom the reader may be drawn and find mutuality. In contrast to the Enlightenment conception of the person, transparent writings reassert the significance of the animal aspects of the psyche: desire, emotion, bodily sensation. Perhaps the most dramatic forms of such writing are found in the newly emerging genre of auto-ethnography. In traditional ethnography the social scientist reports on the lives of others (e.g., the Trobrianders, Thai prostitutes, medical students). Such research does much to *humanize* the subjects of study, but the person of the scientist remains relatively obscure (the silent, knowing one). In auto-ethnography the cloak of obscurity is cast away, and the ethnographer describes his or her own experiences to illuminate a particular way of life. Carolyn Ellis, one of the leading exponents of auto-ethnography, explains in her writing on the mother/daughter relationship that she was "showing the connections among the seasons of a woman's life and encouraging readers to sense what I am feeling as well as hear what I am thinking. And to express their own feelings and think about their own experiences" (Bochner & Ellis, 1996, p. 14).

To illustrate more directly, here is Carol Ronai (1996) detailing aspects of what it is like to be parented by a mentally retarded mother.

> I resent the imperative to pretend that all is normal with my family, an imperative that is enforced by silence, secrecy, and "you don't talk about this to anyone" rhetoric. Our pretense is designed to make events flow smoothly, but it doesn't work. Everyone is plastic and fake around my mother, including me. Why? Because no one has told her to her face that she is retarded. We say we don't want to upset her. I don't think we are ready to deal with her reaction to the truth. . . . Because of (my mother) and because of how the family as a unit has chosen to deal with the problem, I have compartmentalized a whole segment of my life into a lie. (p. 115)

Much auto-ethnography does sustain the modernist tradition of the singular or coherent self. The sense of "you know me as the one who is writing here" is apparent; and yet, living a normal, complex life also means that most of us possess multiple selves; that is, differing ways of talking and being that stand as "me" for different audiences. In this light, adventurous social scientists are drawn to experiment with polyvocality, or the use of multiple genres of self-representation. I was first struck with the power of genre multiplicity when I listened to a presentation by the African American theorist Cornel West. West held me in thrall as he deftly mixed the rhetorics of formal theory, middle-class straight talk, and the argot of the black preacher. Where one voice didn't reach me, another did; the combined force was enormously compelling.

One of the earliest and most provocative adventures in polyvocality is represented in Michael Mulkay's 1985 volume, *The Word and the World*. The work is particularly interesting, as it demonstrates how abstract theory—virtually a private reserve of modernist formalism—can be rendered personal. For example, in the introductory chapter, the voice of a querulous interlocutor is interspersed throughout. The expository Mulkay speaks formally of "extending the range of analytical discourse to include forms not considered appropriate" (p. 10). Mulkay as the impious interlocutor replies, "That sounds very attractive in principle, but it ignores the important distinction between fact and fiction" (p. 10). Mulkay goes on to explain to the interlocutor that even within science "what is fact for one [scientist] is no more than fiction for the other" (p. 11). The interlocutor rebuts,

"Aren't we in danger of confusing two different meanings of 'fiction?'" Later chapters include an exchange of correspondence between Marks and Spencer, letters from these individuals to Mulkay himself, and a discussion among a group of inebriated participants at the Nobel ceremonies.

What has happened here to the traditional criteria of excellence in scientific writing? Somehow, as the personhood of the author expands in dimension, these criteria seem to diminish in importance. For example, there is little demand in polyvocal writings for verbal economy; is it possible that austere writing generates the sense of diminished personhood? Such polyvocal writings are anything but dispassionate; isn't this passion superior to hiding one's investments beneath a misleading cloak of neutrality? Nor is there a strong demand placed on logical coherence; indeed, polyvocal writing stands as a critique of the criterion itself. In such polyvocal writings, clarity and certainty of the traditional variety give way to ambiguity and ambivalence; in reaching for a full relationship through writing, there is no "comprehensive account" because space must always remain for the added voice of the reader.

More important than what is missing in these experiments in written forms is what they create. As I experience them, they seem to strike up a different form of relationship than what I have commonly encountered. Rather than the cold, brittle, and intrusive rationality of the autonomous other, I often find warmth, spontaneity, and the admission of foibles, all of which draw me to the writer. I am not positioned as an ignorant audience by transparent writing; I am not moved into a posture of self-defense. Rather I am invited into a state of shared sub-jectivity. By writing in the fullness of the first person, I as the reader am invited to imagine myself as the writer, to feel and think with the writer. Thus the boundary between author and reader is diminished. Further, with substantial reliance on affectively charged language—discourse of values, desires, emotions, and spirit—I come to experience the writing in a different way; unlike my reaction to traditional writing, I may come to experience my entire body joining with the words. Further, the sense of hierarchy and competition induced by traditional writing also subsides. With reasoned argument, the dimension of superiority/inferiority is always at hand; however, when the author speaks from experience I am likely to participate as an equal. With the admission of foibles (such as personal bias), I am no longer positioned as an inferior; with the expression of multiplicity, I am no longer so protective of my own incoherencies. I am not a competitor in this world of inscription but linked to the writer in the project of inquiry.

I am also struck by the way these iconoclastic adventures affect my sense of sub-cultural boundaries in the social sciences. By writing in a way that gives me the sense of the writer as a full person, concern with bounded disciplines seems thrust aside. The writer first of all seems to be a human being engaged in inquiry; that he or she happens to have a PhD in a given area seems quite secondary. Much the same subversion of disciplinarity occurs when an author moves toward polyvocality or multiple genres. When writers represent themselves as collectivities, how should I, the reader, go about identifying their "true discipline?" And, if I resonate with one or more of their voices, why should I care?

RELATIONAL WRITING: SELF WITH OTHER

> We then no longer need . . . the "endless safety" of ideologies but prize the "needless risk" of acting and interacting.
>
> Victor Turner, *From Ritual to Theater*

> The concrete language of the theater can fascinate and ensnare the organs. It flows into sensibility. It liberates a new lyricism of gesture which, by its precipitation or its amplitude in the air, ends by surpassing the lyricism of words. It ultimately breaks away from the intellectual subjugation of language.
>
> Anton Artuad, *Theater of the Absurd*

Whose words are my fingers on this computer keyboard now tapping into place? Certainly not my own. If they were mine alone, would they even qualify as words? Would they not be nonsense? Is it possible that my words are, then, borrowed from others—and I but a counterfeit personality? As Mikhail Bakhtin might put it, each moment of speaking is a form of ventriloquism—taking words from one source and expressing them elsewhere. But this answer cannot suffice, for we should soon find ourselves spinning into infinite regress. If my words are those of others, where indeed did they acquire them? From others? These words I am now writing, if they are to be meaningful at all, must be born within our relationship. They are not mine, nor yours, but ours. We play the game of language together, but without an ultimate referee, we generate endless variations on themes that are themselves variations without identifiable origin.

If this is so, then what are we to make of these pronouns "I" and "you?" Are they not misleading, creating artificial distance and disconnection? No, we are not one; "one" is itself a defection. But we may pay homage to that primordial process of relationship to which we owe the very possibilities of you, me, and us, without which there would indeed be no sense of the real or the good—no reason for writing at all. This is indeed the ultimate implication of the dialogues on social construction.

How can our forms of writing bring relatedness itself into common consciousness? In hammering out forms of relational representation, can we not bring into being a new scholarly culture? Distance, alienation, competition, and hierarchy might all recede. In their place we might hope for celebrations of communion, invite exploration without fear, and enable a collaborative construction of better worlds. (I fear the words now become inflated, naïvely optimistic, sophomorically idealistic; but then again, if we are to live in meanings of our own making, why not set out for paradise?)

Writing in a relational key now emerges with increasing frequency in the social sciences. Drawing from precedents in the humanities, social scientists experiment, for one, with dialogic writing. Rather than writing as a singular agent controlling the meaning and defending the sanctuary of self, social scientists write with others in a way that subverts the single truth and weaves together disparate strands of discourse to form a complex tapestry of meaning. To illustrate, I offer an excerpt from one such exploration, a *trialogue* in which I (as a scholar) join two practicing therapists, Lynn Hoffman and Harlene Anderson, to speak out against a new movement in managed care in which therapists are to diagnose relational dysfunction (Gergen, Anderson, & Hoffman, 1997). Most of our conversation has been an attack on diagnostics. Here we begin to reflect on our conversation.

> **KJG**: One hope that the three of us shared in this effort, was that the trialogue as a form of writing would itself demonstrate some of the advantages of a constructionist orientation over relational diagnosis. What happens if we depart from monologue (which parallels the singular voice of diagnostic labeling practices) and approach a multivocal conversation (favored by the constructionist)? In some degree I think we have made good on this hope, inasmuch as each of us has brought a unique voice to the table, drawing from different experiences, relationships, and literatures. Our case is richer by virtue of our joint participation. At the same time, because there is so much general agreement among us, the trialogic form hasn't blossomed in fullest degree. We have not yet cashed in on its catalytic potential.
>
> To explore this possibility, I want to focus on a point of disagreement. How can we treat conflict within this conversational space in a way that is different from a monological orientation (where the interlocutor typically shields internal conflicts in favor of achieving full coherence)? The fact is that I do not, in the case of diagnostics, favor Lynn's preference for joining "what is already in place." As she points out, "the process of definition is the primary framing act of any kind of therapy or consultation" and, by virtue of our various critiques, proposes to multiply the range of definitions, even to include those of the clients themselves.
>
> Now, I realize that it is perhaps easier for me to take this strong position because I am not a therapist and do not depend on maintaining the therapeutic traditions for my livelihood. . . .

> **HA**: Ken suggests that our trialogue has not created the catalytic potential that he hoped to achieve. For me, it has created more thoughts than my written words reveal. I have more of a dialogue in my head about diagnosis, and I frequently bring the issues of diagnosis into my conversations with colleagues and students. As in therapy, is the catalytic potential ever visible?

Can our words on paper further the dialogue about diagnosis for others? I hope so. . . .

LH: It does seem that the conversation is now taking us into new spaces. The question I have is whether the shift would have happened if I had not "joined the opposition" or if Ken had not chosen to "disagree"? If we had used a debate format from the outset, with each person taking a different side, could we have reached this point earlier? Catherine Bateson said at a recent conference that to have the kind of improvisational conversation she finds useful, people first have to establish that they have a common code. So perhaps it is a matter of stages. What do the two of you think?

KJG: In response to Harlene's last comments, it seems to me that therapists struggling to find a niche in managed care apparently see no other way out but to stay within the diagnostic framework. Although I have opted out of this framework, I felt that I should put myself back in to represent their side. But I think Harlene is right to say that this shift toward the medical metaphor not only distances us from our customers but makes us less effective. (p. 231)

To be sure, the trialogue has its imperfections. But what inspired me in this effort was the fact that I was able to work together with professional practitioners, subverting the otherwise self-serving binary of pure vs. applied science. By layering our voices in this way, a far more powerful case could be made. And yet, our arguments together admitted their incapacity to grasp the whole, to create "a final word." In effect, we instantiated the argument we were making in the case of diagnosis. Further, I learned from the writing that it was not a matter of articulating a position already held, but of adding complexity to my understanding through interchange. The process itself also helped to generate bonds that continue to be nurturing and productive. I have now introduced dialogic writing projects into several of my classes, and with occasionally stunning results. To write from within the context of ongoing conversation grants a sense of importance to one's contributions; one writes to others who rely on him or her to sustain and expand the discussion. Further, students are freed to employ wide-ranging genres, not only academic formalities but street talk, intimate talk, irony, humor, and more. The composite breathes life into their relationship and a shared sense of excitement as they move into untraveled territories.

One interesting feature of dialogic writing is its consciousness of address. Rather than the impersonal form of address so characteristic of traditional social-science writing (presuming a single knower speaking outward to a faceless community of the unknowing), dialogic writing is for someone in particular; namely, one's interlocutor. In this sense, the writing calls attention to its performative or illocutionary character; we see it more clearly as a constituent of an ongoing social practice. This performative feature may be accentuated in many different forms of discourses; some courageous social scientists experiment, for example, with poetic forms of inscription. While poetic writing is not addressed in the same degree as dialogic writing, it is performed for an audience, typically in such a way as to invite others into a richer and more fully embodied relationship. To illustrate, here feminist scholar Laurel Richardson (1997) reflects on the nature of her own scientific writing.

While I was Writing a Book

my son, the elder, went crazy
my son, the younger, went sad
nixon resigned
the saudis embargoed
rhodesia somethinged
and my dishwasher failed

my sister, the elder, hemorrhaged
my brother didn't speak to me
my ex gurued and overdosed
hemlines fell and rose
texans defeated the e.r.a.
and my oil gaskets leaked

my friend, the newest, grew tumors
my neighbor to the right was shot
cincinnati censured sin
and my dracena plant rotted

I was busy. (pp. 203-204)

A poem often functions to draw readers into a more intimate relationship with the author. What we presume to be deep inside the writer is turned outward for the reader to explore and possibly ingest. The writing invites the recognition of mutuality. In a related vein, a small number of innovative ethnographers now experiment with ways of rendering the words of those under study in poetic form (see, for example, Glesne, 1997). The attempt here is to portray the essences of people's accounts of their lives, but to do so in a way that expresses the feelings that the native evokes in the ethnographer. Often, the ethnographer will weave together the actual words of those "under study" within the poem. The poem, as well, is intended to generate empathic resonance within the reader. In effect, speaker-scholar-reader all move toward a collective subjectivity.

If you scan the dialogic writing illustrated earlier, you also see that it resembles a theatrical script. In a sense the three of us have written a small, though somewhat flat, piece of theater. The performative character of the dialogue becomes more transparent in the case of poetic writing. Here we find that the writing serves as a virtual gift for the audience. From poetics it is but a small step toward envisioning scholarship as full theatrical performance. The relational implications are also compelling. To achieve dramatic effects typically entails far more than words. Strong theater often requires the full coordination of movement, light, sound, objects, and scenery, as well as a rarefied relationship between actors and audience. In comparison to theater, writing is a minimalist relationship. Perhaps the key figure in developing a rationale for considering theater as a vehicle of scholarly expression was Victor Turner (1982). As Turner proposed, ethnographic documentation, including film, fails to communicate much of what it means to be a member of the society filmed. The more adequate mode of understanding would be generated by "turning the more interesting portions of ethnographics into playscripts, then acting them out in class, and finally turning back to ethnographies armed with the understanding that comes from 'getting inside the skin' of members of other cultures" (p. 90). The pedagogical implications of theatrical participation have now become well-developed in performance studies (see, for example, Carlson, 1996). Gay and lesbian groups, in particular, have pressed forward the political potentials of performance (see, for example, Case, Brett, & Foster, 1995).

Stimulated by constructionist challenges to traditional forms of representation, social scientists are increasingly drawn to theatrical forms of representation. To illustrate, I share the work of Mary Gergen (2001) who has attempted to close the gap between social scientist and performance artist. The following is an excerpt from a monologue presented by Mary (complete with red boa) at several social-scientific meetings. Her attempt in the piece is to both critique certain deconstructive tendencies in postmodern scholarship, while conjoining the positive potential of such work with feminist theory and practice. Here "First Woman" speaks with "PM Man."

> **First Woman**: Who are they trying to scare off? Full of Power and Manipulative Control, Abundant Resources, Speed, Complete Management. The New Army . . . pulling the rug out from under the OLD Guard (Didn't we all want to run out of the stands and Cheer!!!) Down with the OLD ORDER . . . Foundations of Modernity, split into Gravity's Rainbow/Rules shredded ribbons adorning the May POLE, wavering in the Breeze of breathtaking words/ ABSOLUTE-ly nothinged by the shock-ing PM tropes/smashing icons with iron(ic)s/Wreaking CON-SENSE with NON-SENSE/
> PARODYING
> PARADING
> PANDERING
> PARADOXING
> PLAYING
> POUNDING
> PRIMPING
> PUMPING
> What fun (singing) London Bridges Falling Down. (Then shouting) DE-CONSTRUCTED (Resume singing) MY FAIR LADY.

Where can We jump in? Shall we twirl your batons. Can we all form a circle/Dance around the fire? The Pole? The falling bridges? Give us a hand . . .

PM Man: All they ever want are handouts . . . Give'em an inch they'll take a mile. How many inches do they think we've got? . . . Besides, can't you see we've got play to do? It's not easy just going off the play each day. . . . It's not something you can just join like that. . . . We're in the wrecking business. What business is that of yours? "You make, we break" . . . Next thing you'll want us to settle down and play house. We've got to be movin' on. . . . Don't call us, we'll call you.

First Woman: That call has a familiar ring to it. The call of the WILD. (p. 188)

Form and content now merge, embodied actor tissued to reader forged to world with writing no longer serving as a conduit for distant minds but now constituting the world itself. Both form and content manifest and create relationship, and from such a relationship the value of life itself springs forth.

Are these movements toward self-transparency and self/other connection altering the landscape of relations in academic culture? The reader's own experience of the present effort may serve as a weathervane. If you found yourself distanced by the writing included in the initial section, you may represent a great number who are eager for cultural change. If the writings of the second section were more appealing, you may be among those creating more caring relations within the scholarly sphere. If the move toward the performative representation was most enticing, you may indeed be participating in a new breed of scholarly relations—both passionate and playful. Anyone who has attended a conference on qualitative research will understand this very well.

CLOSINGS AND OPENINGS

I close with two specific observations on current transformations in scholarly culture. First, I find that scholars who moved in directions favored by social constructionist thought are finding it easier to realize their inclinations in their practices of teaching than in their forms of scholarship. Strong pedagogical movements toward collaborative and paired writing (Ede & Lunsford, 1994; Reagan, Fox, & Bleich, 1994; Topping, 1995) are highly congenial with the dialogic experiments favored by constructionism. Similarly resonant is the increasing emphasis on cohort programming and learning (see Saltiel & Russo, 2001). Such pedagogy may succeed in generating future scholars whose orientation to their colleagues will be far more communal than heretofore (Damrosch, 1995).

Second, with the flowering of new forms of representation, the very idea of bounded scholarly cultures may deteriorate. As scholarly writing is opened to the full range of existing genres, we may hope to see the regulatory effects of the disciplines wither. Innovative forms of writing open the door to the full range of human expression, and as they do so the question of disciplinary identity recedes in significance. Already, substantial numbers in social science are traversing disciplinary boundaries. Conferences and journals on qualitative methods, the rhetoric of science, feminist studies, action research, culture and technology, and discourse and society (to name but a few) are virtually discipline blind. Here we approach the end of rigid regulation in the social sciences and the beginning of—even the distinction between—what is inside and outside the walls of academia. Ideally scholars in the social sciences might join collaboratively with the public more generally in the attempts to increase the understanding and well-being of humankind.

Of equal significance, however, is whether we can anticipate further erosion in disciplinary strictures and boundaries that would carry across the scholarly arena, and most particularly, link the two cultures of academia: the techno/scientific and the humanistic. In large measure, I think the answer to this question will depend on the capacity of more humanist scholars to make social constructionist ideas intelligible and useful within the enclaves of natural science. Once constructionist intelligibilities are appreciated and regulatory practices are weakened, then forms of representation will expand, and relationships will become more flexible. Unfortunately, constructionist ideas have typically been used for purposes of unmasking the authority of the more trusted (and unreflexive) traditions of social science. Indeed, this mounting volume of critique is largely responsible for the science wars. In my view, further crossing of boundaries will require a new and more collaborative posture among humanist/constructionist scholars. Antagonistic scholarship must give way to mutual inquiry. Ironically, such scholarship may have to obey the established rules of representation. Without the conceptual background, the new genres of expression often seem to be mere play. Borrowing from Wittgenstein, it may be necessary to mount the traditional ladders of expression in order to ascend into the domain of freedom. After the climb, we may properly kick the ladders away.

REFERENCES

Bochner, A. P., & Ellis, C. (1996). Talking over ethnography. In C. Ellis & A. P. Bochner (Eds.), *Composing ethnography* (pp. 13-45). Walnut Creek, CA: AltaMira.

Brodkey, L. (1987). *Academic writing as social practice*. Philadelphia, PA: Temple University Press.

Carlson, M. (1996). *Performance: A critical introduction*. London: Routledge.

Case, S., Brett, P., & Foster, S. L. (1995). *Cruising the performative*. Bloomington: Indiana University Press.

Damrosch, D. (1995). *We scholars: Changing the culture of the university.* Cambridge, MA: Harvard University Press.

Ede, L., & Lunsford, A. A. (1994). *Singular texts/plural authors: Perspectives on collaborative writing.* Carbondale: Southern Illinois University Press.

Gergen, K. J. (1994). *Realities and relationships.* Cambridge, MA: Harvard University Press.

Gergen, K. J., Anderson, H., & Hoffman, L. (1997). Is diagnosis a disaster?: A constructionist dialogue. In F. Kaslow (Ed.), *Handbook for relational diagnosis* (pp. 102-118). New York: Wiley.

Gergen, M. M. (2001). *Feminist reconstructions in psychology.* Thousand Oaks, CA: Sage.

Glesne, C. (1997). That rare feeling: Re-presenting research through poetic transcription. *Qualitative Inquiry, 2,* 202-221.

Krippendorf, K. (1993). Conversation or intellectual imperialism in comparing communication (theories). *Communication Theory, 3,* 252-266.

Kuhn, T. (1962). *The structure of scientific revolutions.* Chicago, IL: University of Chicago Press.

Levinson, S. L. (1989). *Pragmatics.* Cambridge, UK: Cambridge University Press.

Mulkay, M. (1985). *The word and the world.* London: George Allen & Unwin.

Reagan, S. B., Fox, T., & Bleich, D. (1994). *Writing with: New directions in collaborative teaching, learning and research.* Albany: SUNY Press.

Richardson, L. (1997). *Fields of play.* New Brunswick, NJ: Rutgers University Press.

Ronai, C. R. (1996). My mother is mentally retarded. In C. Ellis & A. P. Bochner (Eds.), *Composing ethnography* (p. 115). Walnut Creek, CA: AltaMira.

Saltiel, I. M., & Russo, C. S. (2001). *Cohort programming and learning.* Malabar, FL: Krieger.

Shotter, J. (1997). Textual violence in academe: On writing with respect for one's others. In M. Huspek & G. P. Radford (Eds.), *Transgressing discourses: Communication and the voice of the other* (pp. 17-46). Albany: SUNY Press.

Sia, T. L., Lord, C. G., Blessum, K. A., Thomas, J. C., & Lepper, M. R. (1999). Activation of exemplars in the process of assessing social category attitudes. *Journal of Personality and Social Psychology, 76,* 517-532.

Smedslund, J. (1988). *Psycho-logic.* New York: Springer-Verlag.

Topping, K. J. (1995). *Paired reading, spelling and writing: The handbook for teachers and parents.* London: Cassell.

Tracy, K. (1996). *Colloquium: Dilemmas of academic discourse.* Greenwich, CT: Ablex.

Turner, V. (1982). *From ritual to theatre.* New York: Performing Arts Journal Publications.

PART II

Regulation and the Possibilities of Action: Agency, Empowerment, and Power

CHAPTER 7

Shifting Agency:
Agency, *Kairos*, and the
Possibilities of Social Action

Carl G. Herndl and Adela C. Licona

After surveying conceptions of agency in rhetoric and professional communication, we argue that agency is not an attribute of the individual, but the
conjunction of a set of social and subjective relations that constitute the
possibility of action. The rhetorical performance that enacts agency is a form
of *kairos,* that is, social subjects realizing the contextualized opportunities for
action. Drawing on Foucault, Bourdieu, Bordo, and Burke, we contend that
agency is a diffuse and shifting social location in time and space, into and
out of which rhetors move uncertainly. Constrained agency emerges at the
intersection of agentive opportunities and the regulatory power of authority.
These reconcepualizations of agency, authority, and regulation complicate
the framework for investigation and interpretation of how subjects function
in cultural practices that reproduce knowledge, power, change, and identity.

AGENCY REDEFINED:
AN OPPORTUNITY IN SPACE AND TIME

The question of agency in contemporary social and rhetorical theory might best
be seen as a response to the failures of the philosophy of action and its humanist
social actor. In cultural studies, the question of agency is an attempt to theorize
the possibilities of radical, counterhegemonic action, especially in the face of
powerful cultural formations. Here we might think of Gayatri Spivak's (1988)
argument in "Can the Subaltern Speak?" and more recently in *A Critique of
Postcolonial Reason.* In this formulation, agency becomes a question of whether

and how the subaltern can make her voice heard and achieve political legitimacy; that is, how she can (re)constitute her identity and (re)position herself within the public sphere. In rhetorical theory, we might rephrase this as a question of how rhetors effect social change. Or, more generally, how people enter into and effect arguments and debates, recalling that in order to participate in a debate, a speaking subject must first be recognized and able to enter the discussion. This last, rather mundane, formulation of agency moves us away from the interventionist politics of cultural studies, a move we make consciously, if also temporarily and conditionally. This broader formulation of agency also provides a strategic perspective on the theoretical issues at play. That is, as one of us has argued elsewhere (Herndl & Bauer, 2003), while we would like to reserve our theory of agency for the activist political positions with which we align ourselves, an adequate theory of agency must account for rhetorical and cultural action across the political spectrum.

Understanding subjects as discursive selves and understanding discourse as symbolic action—an event—with material and social consequences, we are left to question not only who has the authority to speak and represent, but also what the conditions and opportunities are that allow subjects to act to change or to reproduce social, institutional, and discursive practices. If rhetorical agency is the act of effecting change through discourse, how should we think of agency? What sort of phenomenon is rhetorical agency? Moreover, if agency is associated with change in social or discursive practices, what is its relationship to authority? As a figure constructed within institutional relations of value and power, authority often limits and controls discourse and action. Authority, like the author function in Foucault, often regulates discursive behavior and the creation of meaning; it rarifies discourse, in Foucualt's terms (1972), regulating who can speak and what topics are legitimate subjects of discourse. This institutional relationship between agency and authority leads us to introduce the notion of *constrained agency*. But we also suggest that authority and agency are not always opposing forces within complex institutions. We need a more careful understanding of the interaction between agency and those regulative forces that stabilize institutions and practices. Indeed the regulative power of rhetorical and institutional authority is often interrupted as agentive and authoritative motives or, to use Burke's term, overlap. This leads us to ask, How do agency and authority interact when subjects are authorized to speak against the dominant practices or when their discourse maintains dominant social relations? How, finally, are agency and authority related to the concrete individual? What kind of being is a social agent? How can we think about the subjectivity of an author, of a rhetor, in light of the postmodern critique of subjectivity and identity? In order to frame these questions in terms of rhetorical theory, we take up notions of *kairos* and ethos. Kairos implies the moment in time when speaking and acting is opportune and when this opportunity has important implications for a concept of agency. Ethos implies the authority to speak and act with consequences. Ethos, in this regard, can be understood as a legitimating function for a rhetor or subject. Authority implies legitimacy that validates one's right and ability to speak and act within a given context.

In what follows, we argue that agency is the conjunction of a set of social and subjective relations that constitute the possibility of action. The rhetorical performance that enacts agency is a form of *kairos,* that is, social subjects realizing the possibilities for action presented by the conjuncture of a network of social relations. We reconsider the relationship between agency and authority, identifying authority as both a potential constraint and a potential resource to agency depending upon specific contexts. To explore the complexities of this variable relationship, we examine *Disciplining Feminism: From Social Activism to Academic Discourse,* Ellen Messer-Davidow's (2002) recent analysis of women's studies programs that manifests a contradictory relationship between agency and authority. Next, we examine Jim Henry's (2000) *Writing Workplace Cultures: An Archaeology of Professional Writing* in which he demonstrates that agency and authority are often complementary. Not only do these examples help us understand postmodern agency and its relationship to authority, they allow us to see how the same social subject can occupy different, sometimes contradictory, identities and social spaces. Thus the same person is sometimes an agent of change, sometimes a figure of established authority, and sometimes an ambiguous, even contradictory, combination of both social functions. Through rethinking agency and its necessary relationship to authority, we constitute a theory that explains the way social subjects move between identities and discursive functions, and how we are all articulated in passing and shifting ways to different social spaces and practices. Agency speaks, then, to the possibilities for a subject to enter into a discourse and effect change—even change that might serve to further entrench a dominant social order.

AGENCY: THE EXCAVATION

In framing the question of agency, theorists typically struggle with the dilemma of the postmodern subject and her ability to take purposeful political or social action. This has been an important question across the humanities over the last decade. To make progress on this vexed question, we begin by following Raymond Williams' (1977) example and suggest that we consider "agency" as a concept whose problematic emerges from our disciplinary history. There are a number of places we could begin in examining the disciplinary history of the concept of agency; any one of them could be a synecdoche for a broad theoretical position. In *Discerning the Subject* (1988), Paul Smith surveys the problem of the subject and explores a series of theoretical responses. He concludes that poststructural theories of the subject "tend to foreclose upon the possibility of resistance," and resistance is the scene and enactment of agency for Smith (1988, p. xxxi). When Smith discusses the purpose of his book, he describes his theoretical project as "an attempt to discern the 'subject' and to argue that the human agent exceeds the subject as it is constituted in and by much post-structuralist theory as well as by those discourses against which poststructuralist theory claims to pose itself" (p. xxx).

More specifically and somewhat closer to home in rhetorical theory, John Clifford (1991) articulates the outline of the debate in "The Subject in Discourse," where he balances the naïve humanism of *The Little, Brown Handbook* (Clifford's synechdoche for the profession) against his own description of the poststructural subject interpellated by discourse. We take Clifford's formulation to represent the standard understanding of the problem. As Clifford describes it, *The Little, Brown Handbook* suggests that an adequate, well-designed argument will succeed in persuading its audience. Both the writer and the audience in this scenario operate in some version of Habermas's (1970) ideal speech situation as autonomous, rational actors. Against this, Clifford sets out the carceral model of subjectivity, as a nonporous, inflexible category into which subjects are inter-pellated by ideology and determined by discourse. Clifford's position combines Althusser's (1971) ideological interpellation and the most heavy-handed inter-pretation of Foucault's (1972) rules for the formation of statements and subject positions discussed in *The Archaeology of Knowledge and the Discourse on Language*. The critique of the rational, Enlightenment rhetor and the subsequent analysis of structure and determination run through much current work in rhetoric and professional communication. Bernadette Longo's *Spurious Coin: A History of Science, Management, and Technical Writing* (2000), for example, articulates the historical case of this sort of determinism in the development of technical writing. In addition, Brenton Faber's *Community Action and Organizational Change: Image, Narrative, Identity* (2002), explores the power of structural constraints in a number of contemporary examples.

The opposition we have just sketched erases the many differences in emphasis within these general positions and hardens them for schematic clarity. Nonethe-less, we think that this opposition adequately describes the two contending positions with which many of us are all too familiar. The contemporary theory of agency is a response to the theoretical stalemate this opposition represents. Our disciplinary concept of agency has emerged from our feeling that neither of these formulations explains the rhetorical and social phenomena we experience. This situation in which a theoretical stalemate generates a new concept is not unique to rhetorical theory or to debates about the humanist individual versus the poststructural subject. In *Hegemony and Socialist Strategy: Towards a Radical Democratic Politics* (1995), Ernesto Laclau and Chantal Mouffe his-toricize the concept of hegemony as a feature of Marxist theory. Orthodox Marxism had described the historical necessity of the Proletariat revolution, the coming into class consciousness of a unified Proletariat as the subject of history. Unfortunately, that did not happen, and Marxist theory had to explain the lapse. As Laclau and Mouffe write, "[t]he concept of hegemony did not emerge to define a new type of relation in its specific identity, but to fill a hiatus in the chain of historical necessity" (p. 7). We suggest that the theory of rhetorical agency, which is prominent in rhetorical studies, fills the hiatus created by the failure of poststructuralism that Smith (1988) details and rhetoric's wise refusal to recuperate a romantic concept of the individual. That is, the concept of agency does not describe a new phenomenon. Rather, it is necessitated by the failure of poststructural theories of the subject to explain social change or rhetorical action.

If we define agency as Anthony Giddens (1984) does, as not "the intentions people have in doing things but to their capability of doing those things in the first place" (p. 9), then agency looks a lot like power, both as a theoretical notion and as a practical occurrence. To reconceptualize agency, then, as a truly social phenomenon, we turn to Foucault's suggestive remarks on power. In taking up the question of power, Foucault (1982) avoids asking what power is or why it is used. Instead, he asks us to consider the how of power. He writes, "[t]o put it bluntly, I would say that to begin with the analysis of a 'how' is to suggest that power as such does not exist," (p. 785). When critics talk about the what and why of power, they reify a set of social relations. When this happens, Foucault continues, "an extremely complex configuration of realities is allowed to escape when one treads endlessly in the double question: What is power? And Where does power come from?" (p. 785). By contrast, asking how power emerges and circulates leads us to investigate what Foucault terms the "thematics of power." Following Foucault, we suggest that, like power, agency as such, does not exist. As a general rule of thumb we suggest that scholars mystify social reality whenever they use agency after a transitive verb. Agency cannot be seized, assumed, claimed, had, possessed, or any of the many synonyms for these transitive verbs. As Susan Bordo (1998) has argued of power, agency "is in fact not 'held' at all; rather people and groups are positioned differentially within it" (p. 1107). The reality of agency is a question of positioning within what Bordo, quoting Foucault, describes as the "multiple 'processes, of different origin and scattered location,' regulating and normalizing the most intimate and minute elements of the construction of time, place, desire, embodiment" (p. 1107). Bordo's notion of the differential positioning of subjects within relations of power leads back to Foucault's (1982) revised ontology of power. Power is not a transcendent thing. Foucault continues, *"[p]ower exists only when it is put into action,* even if, of course, it is integrated into a disparate field of possibilities brought to bear on permanent structures. This also means that power is not a function of consent," (p. 788, emphasis our own). Throughout his work, Foucault argues against the sovereign model of power. Centralized and total, sovereign power is an attribute of an individual. If we understand power as a set of relations, however, it no longer requires that we connect it to an autonomous individual. So, too, with agency. It does not reside in a set of objective rhetorical abilities of a rhetor, or even her past accomplishments. Rather, agency exists at the intersection of a network of semiotic, material, and yes, intentional elements and relational practices.

If we define agency as self-conscious action that effects change in the social world, then agency is contingent on a matrix of material and social conditions. It is diffuse and shifting. In contrast to the implied model of agency as an attribute or possession of individuals, agency is a social location and opportunity into and out of which rhetors, even postmodern subjects, move. Recently Radha Hegde (1998) has referred to agency as "the coming together of subjectivity and the potential for action" (p. 288). Further, in *The Logic of Practice*, Pierre Bourdieu's (1990) theory of practice defines agency as the conjunction of a subject's habitus and the changing social conditions for action. Habitus, according to Bourdieu, is the set of durable dispositions inculcated in the subject by her past experiences, and these dispositions adjust the subject's rhetorical actions to the continually changing situation (p. 53). From this perspective, rhetorical action is neither determined by structures nor the domain of the autonomous individual. It is the conjunction of the subject's dispositions and the temporary and contingent conditions of possibility for rhetorical action that begin to define what we term an *agent function*. This understanding of agency articulates the poststructural subject to the radical contextualization of cultural studies.

Like the author function that Foucault identified, the agent function implies agency before the agent. But what we call the agent function is relatively less stable than Foucault's author function, and it reveals the moment or opportunity when and where action is possible. Bourdieu (1977) refers to this temporal and spatial overlap when he speaks of that which allows for action and the agency function to be engaged. His notion of the interval articulates the dimension of time to place. Bourdieu illustrates this point quoting Leach (1962) who refers to "an area where the individual is free to make choices so as to manipulate the system to his advantage" (quoted in Bourdieu, p. 53). For purposes of our own theory building we see the articulation of the materiality of time to place and practice as an important and necessary move (see also Condit, 1999, p. 176). The place from which one speaks or writes and within which one acts is a social space, but one that exists in time.

Before we can develop an adequate theory of agency, however, we must make three radical interventions. First, we must sever the metonymic identity of agent and agency, as well as the metonymic identity of author with authority. Second, we must reverse the order in which we think of these relationships; agency phenomenologically precedes the agent and authority phenomenologically precedes the author. In so doing, we differentiate the subject from the social agent as well as the subject from the author. Specifically, we contextualize agents and authors as sites of an agency function and an authority function. To borrow Karlyn Kohrs Campbells' (2003) term, the agent is a "point of articulation" rather than an origin. Third, we must reveal the necessary, if also shifting, relationship between agency and authority. Throughout these interventions we reconsider the implications of power and representation. Our efforts are undertaken to offer a more complicated framework for interpretation to those researchers of technical and professional communication in both academic and nonacademic settings who are investigating how subjects as writers function in the cultural practices which (re)produce knowledge, power, and identity. Our use of the parenthesis in the preceding sentence to mark the ambiguity with which discourse relates to knowledge and power suggests both the ubiquity of institutional regulation of rhetorical action and the opportunities, however fleeting, to escape regulation and generate change.

METONYMY INTERRUPTED:
AGENCY AS SOCIAL PRACTICE

Not only is agency an artifact of our disciplinary history, the concept is compromised by the history from which it emerges and to which it is a response. More specifically, the stalemate between poststructural and romantic traditions obscures the metonymic identification of agency with agent and mires the concept of agency in a specific theoretical legacy. Despite the postmodern theory of the subject, agency continues to be thought of in terms of the individual. Theorists interested in promoting social change write of someone "having" agency, of agency being the attribute of an individual speaker or writer. So Lisa Ede, Cheryl Glenn, and Andrea Lunsford (1995) write, for example, that someone can "claim authority and agency" (p. 423). Similarly, in his discussion of rhetoric and change, Brenton Faber (2002) writes that "[a]lthough these people can claim some degree of agency, they still cannot claim to be completely free of social structures" (p. 121). Faber is careful to articulate a theory of power in connection with social change, but the vestigial language of the humanist individual nonetheless haunts his discussion of agency. We think that this is a ubiquitous but fundamental error and one that misdirects our theory building. The examples we have used to describe the reification of agency and the stabilization of the agent as a substantive category come from Marxist and feminist work that we support—work that is aimed at changing social relations and making them more equitable. But we worry that, like much work in cultural studies that celebrates the way subcultures resist dominant formations, this theory of agency reifies the individual. Furthermore, Meagan Morris (1990) has argued that political and ideological resistance has become "banal" in much cultural studies, all countercultural forms being seen as careful resistance and struggle (p. 14). Similarly, we worry that radical subjects seize agency and become social agents all too regularly in our theoretical discourse. Theories of agency are typically laudable attempts to catalyze action and social change, but we worry that they are more epideictic than analytic. We share the sense of purpose that motivates theories of agency, but worry that they produce an inaccurate notion of agency that reifies autonomous agents and obscures the network of material and textual conditions upon which agency depends. If we are correct and these theories are really epideictic, they perform a kind of "strategic essentialism," in

Spivak's (1990) or Emma Perez's (1998, p. 87) terms, a necessary move at times, but a potentially misleading one.

To uncouple the agent/agency metonymy, we begin with Delip Goankar's (1997) critique of recent scholarship in the rhetoric of science. Indeed, it is an indictment of the field itself. Goankar begins the critical work of dismantling the figure, which collapses agency into the agent. Goankar argues that the humanist model of persuasion locates agency within the individual and in his or her conscious intention. Goankar writes that "[t]he agency of rhetoric is always reducible to the conscious and strategic thinking of the rhetor. The dialectic between text and context, a topic of considerable interest today, is already pre-figured in the rhetor's desires and designs. Such is the model of intentional persuasion, still dominant, but under trial" (p. 49). What Goankar calls "strategic thinking"—the notion that a specific scientist managed to win a scientific debate through intentional use of rhetorical skill—"marginalizes structures that govern human agency: language, the unconscious, and capital" (p. 51). This strategy, which is used to explain scientific argument and social change, has a powerful tradition. For example, Hayden White (1990) characterizes Darwin's success as his ability to combine the tropic form of metonymy with the "conservative" (p. 129) and "more comforting" (p. 133) tropic form of synecdoche. Whether or not he consciously intended to do so, White does not venture an opinion here on whether Darwin's rhetorical formulation is responsible for the success of his argument. White's analysis does not explicitly attribute strategic thinking to Darwin, but it does leave untroubled the humanist assumptions that Goankar critiques. We see Goankar's analysis as a critique of the metonymic identification of the individual agent with agency. Despite the powerful and widely accepted poststructural critique of the individual, we think that theories of agency remain haunted by a hangover from romantic voluntarism. Like hegemony in orthodox Marxism, agency is theorized as a mediating position, compromised by the way it is embedded in the metaphysics of the individual. Its postmodern pedigree notwithstanding, agency is still hampered by the vestiges of humanist models of action. This disciplinary legacy leads, we think, to two mistakes. Agency is tied to the concrete individual, whether she is figured as the *individual* or as the *subject*. Second, agency is theorized as a thing, something agents *have, possess,* or *gain.*

A REVERSAL:
AGENCY BEFORE THE AGENT/AUTHORITY
BEFORE THE AUTHOR

Once we cease thinking of the agent as the origin or locus of agency, we can make a second, more radical move. If we read the agent/agency metonymy backwards and consider the move from agency to agent, we can argue that it is the social phenomenon of agency that brings the agent into being. We argue that social agents only exist when subjects occupy a set of relations and their agentive possibilities in a certain way. As Sullivan and Porter (1993) have argued, a theory of agency need not evacuate the category of the subject or see the subject as a coherent, unified being. Rather an adequate theory of agency "proceeds on the basis of multiple and shifting subjectivities that enable opportunities for change, at least at local levels" (p. 42). Thus "the human agent," as Smith (1988) argued, "exceeds the subject" (p. xxx).

The postmodern subject becomes an agent when she occupies the agentive intersection of the semiotic and the material through a rhetorical performance. Agency here does not reside in the individual, and this conception does not deny the power of language, (con)textuality, the unconscious, and capital. Agency is a social/semiotic intersection that offers only a potential for action, an opportunity. Subjects occupy that location skillfully; a rhetor's abilities and accomplishments make a difference in how her performance is accepted. While the performance itself is not adequate to constitute agency, no matter how often it is repeated, it is part of the complex relations that make agency possible. In making this argument that the enactment of agency brings the agent into being as agent, we are moving from Foucault's insight that discourse creates the objects about which it presumes to speak, to Judith Butler's (1990) argument that the performative constitutes the subjectivity of the performer. Foucault's now commonplace insight in *The Archaeology of Knowledge* (1972) is that powerful discourses constitute the objects about which they speak as they create knowledge. Further, Butler has argued that the performance of gender constitutes the gendered subject itself through the performance. Whereas Butler denies the existence of anything prior to the performance ("a doer behind the deed," p. 142), however, we argue that postmodern subjects—split, contingent, driven by desires and multiply inter-pellated—exist before or outside their agentive or authoritative performances, outside the shifting social location of agency and authority. This subject-in-motion resembles Lawrence Grossberg's (1992) "nomad." Grossberg writes that "the affective individual always moves along different vectors . . . like the nomad, it carries its historical maps (and its places) with it, its course is determined by social, cultural, and historical knowledges but its particular mobilities *are never entirely directed or guaranteed,*" (p. 126, emphasis our own). Like Grossberg's affective individual who is articulated into distinct ideological identities, the subject temporarily occupies different agentive and authoritative spaces. But the subject's ability to seize the potential for action is never guaranteed or permanent. The subject becomes an agent when she is articulated into the agent function. Foucault argued that the author function arose in literary discourse as both a principle of thrift, which limited the range of meaning attributable to a text, and as a response to historic changes in the materiality of the text, specifically the rise of copyright laws. Like Foucault's author function, the agent function arises from the intersection of material, (con)textual, and ideological conditions and practices.

BURKE'S DRAMATIST PENTAD:
SYMBIOSIS AND EQUILIBRIUM IN AGENT
AND AUTHOR FUNCTIONS

Foucault's work on power has spurred countless commentaries, and we have made much of it here, but we want to turn to Kenneth Burke's work on the pentad as a useful way for thinking about agency in rhetorical terms. In the current context, Burke's fundamental point in *A Grammar of Motives* (1969) is that rhetorical events result from a complex relation of elements, no one of which is primary. We think that Burke's metaphor of motive addresses the central issue that a postmodern theory of agency must explain: what is the motive force behind rhetorical events? What relationships constitute the necessary conditions and drive behind successful rhetorical acts? When we talk about agency as if it were an attribute of an individual or something a rhetor can *have*, we privilege one of the five moments on Burke's pentad. We impose one ratio on social reality. If the other four points of the pentad are valid analytic tools, as we think they are, then Burke's theory suggests that agency (in our terms rather than his) is the conjunction of the five elements of the pentad. Agency is the conjunction of all the ratios in a rhetorical context. As Burke says, some rhetorical events depend more on one ratio than on others. But we believe that all the ratios are relevant to all rhetorical events even if they are not the dominant element.

AUTHORITY AND THE AUTHOR FUNCTION REVISITED:
ARTICULATING SPACE TO TIME, CONTEXT,
AND PRACTICE

Like agency, authority is a social location, (re)produced by a set of rela- tional practices. The authority to speak—a speaker's authority in discourses and debates—is a social identity that is occupied by a concrete individual but emerges from a set of social practices. In this sense, authority is tied to classical notions of ethos. Authority is (re)produced by the authority function, and it legitimizes a subject to speak and act for or against change. The authority function lies outside the subject and is therefore not about an individual author/rhetor but instead about the capacity and opportunity to rarify discourse and action. We believe, as does Grossberg (1992), that "authority is not constituted from the identity of the actor but from the already invested worthiness of the site" (p. 381). That is to say, postmodern subjects are complicatedly situated within structures that, at least in part, define the context in which they participate in the author function or the agent function. Authority, like agency, exceeds the subject: it comes before and outside the subject. More specifically, as Grossberg (1992) notes, "authority is not located in the leaders in the community, but in the place that has been constructed, through cultural and intellectual labor, as authoritative" (p. 383). Social practice, context, and space constitute a place in which agency is enacted.

But this place is temporal as well. A "place in time," to use Steven Mailloux's (2003) phrase, where the material and the temporal, combine to constitute the possibility of agency and authority.

Because the authority function reflects the cultural and relational practices that constitute value and power, authority tends to stabilize and maintain the structures within which it is constructed. Social structures tend to be stingy with the social places that took so much work to constitute and in which social groups have so much invested, so much productive past. As it constitutes materiality of authority, the recognizable social place of authority, the authority function is part of what Giddens describes as the recursive and regulative relationship between structure and practice (see Giddens, 1984). Like the author function Foucault described, authority often acts as a principle of thrift, constraining discourse and action and maintaining social practices. In more regulated and institutionalized contexts— the academy and some workplaces we will examine—authority is a prime motive for rhetorical action. In these institutionalized settings, authoritative practices often reveal a power to stabilize, limit, and control meaning and action. Because it authorizes a rhetor to speak, act, and represent, the authority function often represents and reproduces dominant rhetorical and social relations. As it limits the proliferation of meaning and action, authority can constrain agency.

While the authority function often constrains agency, it also rarifies subjectivity within discursive fields. Foucault (1972) argues that as authoritative practices legitimize some speakers, they also exclude other speakers from a given discourse. He notes how authority serves to "impos[e] a certain number of rules upon those individuals who employ it, thus denying access to everyone else. This amounts to a rarefaction among speaking subjects: none may enter into discourse on a specific subject unless he has satisfied certain conditions or if he is not, from the outset, qualified to do so" (pp. 224-225). This process of rarefaction explains how nondominant subjects are all too often excluded from the public sphere because they are not authorized to speak and represent. In other words, authoritative practices can so condition the opportunity for agentive action that it becomes extremely difficult for some subjects (typically those from nondominant groups) to successfully occupy and engage the agentive space.

AGENCY AND AUTHORITY: A RELATIONSHIP OF CONTRADICTION

In *Disciplining Feminism: From Social Activism to Academic Discourse,* Ellen Messer-Davidow (2002) chronicles the emergence of feminist thought and action in the academy over the last four decades. While feminism has successfully established itself within the academy, as evidenced by the proliferation of women's studies programs, Messer-Davidow argues that ultimately feminisms and the *new* knowledges they (re)created have been domesticated by and commodified within the academy and therefore have lost some of their initial radical momentum and agentive potential. We contend that this is not inherent in feminism itself, but that authority as a function within the academy is implicated in the domestication and commodification Messer-Davidow describes.

In her review of the interventionist tactics and relational practices engaged in by 1960s feminists committed to proliferating women's studies in the academy, Messer-Davidow identifies the agentive opportunities that made the discipline of women's studies possible in the academy. The agent function is materialized in Messer-Davidow's work as those practices that agitated for the inclusion of women's studies in the academy. Specifically, she recounts the history of the Coalition of Campus Women (CCW) as informed by the politics and practices of the Industrial Areas Foundation (IAF) and its leader, Saul Alinsky. Using tactics learned from the IAF, the CCW undertook a campaign of direct action, agitating for a women's studies program to include library funding and, among other things, access to administration and administrative records (p. 7). Messer-Davidow notes that to "challenge the academy's knowledge discourse, feminists had to navigate the preliminaries: they had to gain admission to a discipline and win the credentials that authorized them to operate in it" (p. 48).

Messer-Davidow investigates the structures and practices that transformed and disciplined academic feminism. She demonstrates how once women's studies emerged as a legitimate field of study—a discipline—within the academy, it engaged in authority functions that constrained its own agency as a site of alternative knowledges and practices. That is to say, once women's studies was accepted as a discipline, it was transformed "by the [very] structure it had set out to transform" (p. 13); namely, the academy. Women's studies programs have to participate in the power dynamics of the academy, and they do so at the cost of a loss of agency. As Foucault (1972) has pointed out, a discipline exercises a limiting function: "Disciplines constitute a system of control in the production of discourse, fixing its limits through the action of an identity taking the form of a permanent reactivation of the rules" (p. 224). Any academic discipline survives through a struggle between competing demands for innovation and consolidation. Disciplines produce new knowledge, but within and against disciplinary expectations, styles, and policing practices such as hiring, annual reviews, tenure decisions, editorial policies, and curricular structures.

Throughout her work, Messer-Davidow explores the processes and practices that led to the domestication of feminism within the academy, noting that many feminists believe that such domestication is inevitable. She (2002) cites the directors of two university presses who consider that "feminist scholarship, once revolutionary, [has] become repetitious" (p. 205). The repetition of agentive practices institutionalize and legitimize women's studies program, but they also contain their radical potential. The agentive possibilities of radical feminism are also limited by women's studies' relations to other disciplines and programs within universities. In order to maintain its authority within the university, a women's studies program or department must perform the duties and functions commensurate with its institutional status. As programs offer degrees and cross-list courses, they are compelled to engage the prevailing expectations and standards for courses, degrees, and departmental management. Faculty who must face college- and university-level tenure boards are expected to produce recognized styles of scholarship and compete with faculty from other, less politically engaged departments for tenure and research dollars. Thus, the authority function operates both within the discipline and in the general academic field. As Bourdieu might put it, the symbolic capital that women's studies needs to survive in the academy is recognized in a field controlled by the broader academic community and in which radical community and political activism has relatively little value.

Messer-Davidow's work demonstrates two important things for our analysis of agency and its relation to authority. Women's studies programs constituted a radical intervention in the academy when they began to appear. The backlash against women's studies and the attacks by right wing critics (e.g., Bloom, 1987; D'Sousa 1991; Sommers, 1994) attest to the reality of their intervention and revolutionary work. But the necessity to consolidate their symbolic capital within the university and to conform in varying degrees to academic expectations, constrains their agentive possibilities. Thus, we introduce the notion of constrained agency that emerges as a result of the relationship between agentive opportunities and the regulatory power of authority. In this case, authority operates as one point in a revised version of Burke's pentad, producing an agency/authority ratio that defines the motive for radical feminism within the university. Both agency and authority are generated by material practices established within institutional contexts. Where Foucault had defined the author function as a principle of thrift, which limits the dangerous proliferation of meaning, the authority function within disciplines and universities similarly limits destabilizing and potentially dangerous practices, thus constraining the possibilities for rhetorical action.

ARTICULATING AGENCY AND AUTHORITY: OVERLAP, AMBIGUITY, AND SLIP-SLIDING AWAY

The example of women's studies in the academy suggests an opposition between agency and authority—a relation that only (re)produces constraint. As powerful as this constraint can be, however, we want to reconsider the diverse potentials in the relationship between agency and authority. Like agency, authority is realized in contextualized relational practices that define the subject's capacity and the opportunity to function as an authority or an agent. We are particularly interested, however, in the slippage between the agent function (as agency) and the author function (as authority). We want to complicate the tidy opposition Messer-Davidow's work suggests. In complicating the agency/authority ratio, we find it useful to reconsider the movement between these two functions in much the same way that de Certeau (1984) conceptualizes and distinguishes tactics from strategies. De Certeau establishes his now popular distinction between tactics and strategy as the modes of social action associated with resistant and dominant cultural interests respectively. In de Certeau's theory, dominant ideological positions control the social terrain and engage in long-range, stable strategic action, while resistant positions engage in tactical hit-and-run action. While we find de Certeau's distinction falsely dichotomous, we look beyond this dichotomy to the space of overlap between tactics and strategies and between agency and authority. It is this space of overlap that is ripe for reconsideration as a productive and generative space of action and representation revealing both agentive and authoritative relational practices.

De Certeau's understanding of action in everyday life suggests a final element to our theory of agency that is a useful addition to our theory here. De Certeau writes that tactics "takes advantage of 'opportunities' and depends on them. . . . [Indeed tactics] must accept the chance offerings of the moment, and seize on the wing the possibilities that offer themselves [and] . . . make use of the cracks that *particular conjunctions* open in the surveillance of the proprietary powers" (p. 37; italics added). We find the distinction useful in two ways. First, as a discursive distinction, it allows us to talk about the multiply situated subject who engages in both tactics and strategies. More importantly, we are interested in the overlaps between the two and consider them spaces in which agency and authority potentially overlap. We call this motion the slippage between the two and think it allows us to discuss the ways in which subjects are multiply situated and differently able and authorized to speak, act, and intervene. While we understand the distinctions de Certeau makes between tactics and strategies to be primarily discursive, we appreciate the ways in which this distinction allows us to reconsider the implications of time and timing in relational practices that reveal constrained agency and authority. This mobility across space and time is an important part of authoritative acts, agentive opportunities, and relational practices.

The "particular conjunctions" de Certeau (1984, p. 37) identifies are temporal—they are moments of possibility. We do not like de Certeau's sense of an endless guerrilla warfare or his categorical distinction between strategy and tactics; but he is correct when he says that tactics "seize on the wing the possibilities that offer themselves at a given moment" (p. 37). If we think of social space as changing through time in this way, then a postmodern theory of agency returns us to the classical rhetorical notion of kairos. In John Poulakos' (1983) interpretation of Plato's dialogue, *Protagoras*, kairos gets recognized as the "power of the opportune moment" (p. 40). In Carter's (1988) reading of Plato's *Gorgias*, kairos is "a way of seizing the opportunity of the moment in improvisational speaking" (p. 104). Both conceptions of kairos shift away from the individual and toward the opportunity itself—toward the social conjunction or what we call a moment in social space and time. We would push this theoretical formulation of kairos further and read agency as the momentary conjunction of multiple material, semiotic, and intentional conditions of possibility. Lawrence Grossberg (1992) captures the situation nicely. He argues that agency can only be seen when we examine nonepistemological relations of power that are independent of individual actors or groups. Thus "there can be no universal theory of agency; agency can only be described in its contextual enactments. Agency is never transcendent; it always exists in the differential and competing relations among the historical forces at play . . ." (p. 123).

Rhetors, even conceived as postmodern subjects, move into and out of agentive spaces as the result of the kairotic collocation of multiple relations and conditions. Agency, to return briefly to Foucault's (1982) language, is an "extremely complex configuration of realities" (pp. 785-786). As we have moved through de Certeau's distinction between strategies and tactics and confronted these complex realities, our rhetorical sense of kairos further reveals for us the necessity of articulating the materiality of time to discursive, political, social, and cultural practices (see Condit, 1997). For a moment then, we collapse the distinction between tactics and strategy in time and space, acknowledging the multiplicity of ways in which subjects write in the world. In rendering the spaces of overlap visible, we are unearthing the generative, creative, and productive, if also potentially contradictory, space in which multiply situated subjects move. Our theory complicates the practices engaged in by subjects in a given context so that the space of contradiction described in Messer-Davidow's example can be seen as a site of potential productivity and can provide insight into the complexities of subjectivity in organizational or institutional settings.

AGENCY AND AUTHORITY:
A RELATIONSHIP OF COMPLEMENTARITY

Jim Henry's (2000) analysis of workplace culture and practices offers insight into a more ambiguous relationship between agency and authority. Henry is particularly interested in notions of authorship that circulate in the workplace. He explores the way the author function plays out in collaborative writing practices and often marginalizes the work of professional writers. In the current context, however, his research demonstrates the complementary relationship between the agent function and the authority function. Within his study we can identify subjects' movement between agency and authority in ways that blur the boundary between the two. Movement between constrained agency and authority makes visible the shifting and, at times, contradictory positions in which subjects are situated over time and across space. Henry uses articulation theory to identify the relational practices and potential of agency and authority. Specifically, he states that "communication always articulates all subjects implicated in the process to power and knowledge, albeit in varying ways depending upon institutional location, [and] local practices" (p. 143). Henry refers to the development of acuity and expertise (ethos) as a function of authority. The knowledge of institutional or organizational practices offers subjects the opportunity for authoritative and agentive practices and relationships. That is to say, both agency and authority can be constrained by discursive structures but not completely. Opportunities then are represented by the moments when the agent function or author function (re)produce the practices for the subject to speak with authority and act with a potential for change.

Using a number of analytic frameworks, Henry describes the ways the multiply situated discursive self participates in both (re)authorizing practices and agentive intervention. His work identifies the complicated and sometimes nuanced ways in which subjects participate in these (re)authorizing practices with agentive potential. Specifically, in his discussion regarding writers situated in private business and corporations, he eschews the notion of passive agents. Instead, Henry reconsiders the implications of those writers who can claim authority as those "who learn the norms governing representation up and down the organization . . . [who] may be more politically astute in manifesting their ideological positions on organizational products and claiming their share of authority in organizational life" (p. 56). In Henry's discussion of the dynamics between structures and subjects, he describes writers' subjectivities during the acquisition of necessary skills for organizational authorship. In this discussion he recognizes how "editing can be construed as cultural renewal and/or reproduction, a practice with more far-reaching implications than generally acknowledged" (p. 74). Henry uses this example of editor as agent and author as a demonstration of how subjects can participate in authorizing practices with agentive potential.

Henry also investigates technical writing practices and contexts. He discusses technical-document processes, reconsidering the prospects for innovation potentially realized in practices of revision. His discussion acknowledges that technical-document processes, as reflected in "hierarchical and cross-divisional processes that represent culture to its members and to outsiders, may be repetitive and fairly standardized in routing across organizational structures, yet as processes bound up in organizational dynamics influenced by players in various positions, they offer occasions for reshaping parts of these collective procedures" (p. 80). Henry offers a number of observations that reveal how writers in the workplace move or slip between practices that reveal the author function and those that reveal the agent function. He writes that writers' subjectivity is powerfully shaped by collaborative writing practices. He notes that "[u]nder such conditions, they may shape their subjectivities through articulating roles in the organization to roles in document processes; similarly, writers may shape organizational procedures somewhat by bringing skills in organizational analysis to daily activities such as document processing" (p. 81).

Henry describes subjects as discursive selves who live within subjectivities that shift as a function of the collaborative practices. In his discussion of cultural stabilization and change, Henry identifies authors as "agents of stabilization of culture" and also agents of change depending upon the "coincidence between the proposed changes and the organization's goals and underlying discourses" (p. 86). The "coincidence" here reiterates our sense of agency as a temporal-spatial conjuncture of elements. In the context of Henry's research, it articulates authority as a crucial element in agentive action.

Finally, Henry calls for a shift in composition and professional-writing pedagogy and writing practices to reflect the material circumstances and complexities of the multiply situated subject. "Intervening in cultural production and reproduction entails equipping professional writers for such positioning intellectually in classrooms and collaboratively with them later in the ongoing processes and products in which each 'I' becomes implicated" (p. 166). Specifically, he relates a story of one professional writer who was responsible for clipping daily news articles for the organization with which she was affiliated. This writer was positioned such that she was able to participate in both the author and the agent function as we have discussed them (p. 175). This writer engaged in the authoritative practice of deciding which news articles to clip. In practice, she was authorized to recognize which clippings could be deemed relevant to the organization for which she worked. Henry (2000) recognizes the "great potential" of her intervention in the reproduction of organizational practices and culture (p. 175). The agent function was engaged each time the authority function was enacted to intervene in organizational practices and representations from her professional space.

The examples we have selected from Henry's study reveal the space of overlap between the author and agent function we have identified. Several of the researchers involved in Henry's study discuss the ways in which split subjectivity informs the author and agent function so that writers can be positioned to act on behalf of an organization and on their own behalf or on behalf of their collaborative writing partners. Henry notes how authors are those responsible for the text and for the values and cultural norms it adheres to (p. 20). His work uncovers how those cultural norms and values can be altered over time, space, and practice as well.

TAKING A COUNTERCULTURAL TURN

At the beginning of this chapter, we suggested that the question of agency emerged in rhetoric and professional communication when the postmodern critique denied us recourse to the enlightenment individual of liberal ideology, yet failed to replace that individual with a potent social actor. That is a coherent and, we think, persuasive theoretical narrative. But agency also became a central concern in the field of rhetoric when the range of contexts and speaking or writing subjects we considered changed. Research on writing in professional and nonacademic settings confronted us with a wider range of contexts, purposes, and writers than had populated earlier studies of classroom composition. The emergence of cultural studies in rhetoric vastly expanded this tendency. Cultural studies focused attention on ideology and the ways people struggle against domination and attempt to change social practices. This aspect of cultural studies leads us to consider not only how to produce efficacious professional communication, but also to explore the way institutions regulate and normalize discourse and identity. It made the experience of nondominant subjects in discourse and society a central issue for analysis. We suggest that as rhetoric turned its attention more and more on the rhetoric and cultural practices of nondominant and subaltern subjects, many of the assumptions about identity and action that had been invisible in traditional analyses erupted into view. The study of the good man speaking well obscures the issues of identity, power, and the material conditions that support efficacious speech. Writing as a cultural practice performed in a heterogeneous and conflicted space populated by difference and distinction vastly complicates notions of rhetorical action. Agency is the name we give to this rearticulation of cultural rhetoric. We think that the theory we have proposed here helps dismantle the obfuscating tendency within our disciplinary history so that we can move beyond merely acknowledging multiple and shifting subjectivities to understanding the distinct ways in which we get things done in the world.

We have demonstrated the ways in which we can identify the agent function at work in the intersection of a network of semiotic, material, and intentional elements and relational practices. But because our theory begins with the understanding of the postmodern subject as multiply situated and shifting, we think it important to consider how subjects are differently enabled or constrained to act and speak within a given context. The complex and situational relationship between agency and authority suggests not only that social subjects are articulated to contingent agentive spaces, but also that those spaces are often ambiguous and contradictory. The overlap between agency and authority we have tried to outline is just such an ambiguous space. This ambiguity is widespread and it can be productive. To invoke Burke (1969) again, "it is in the area of ambiguity that transformations take place; in fact without such areas transformations would be impossible" (p. xix; see also Sandoval, 2000). Ambiguity offers a potentially creative and generative space for postmodern subjects articulated into social spaces constituted through both the agency and authority functions. We believe that understanding agency and authority as social functions into which subjects are multiply articulated brings the theoretical understanding of subjectivity that emerges from cultural studies closer to the textured and textual understanding of specific rhetorical practices. It helps explain how subjects get things done in ways that collapse any easy opposition between agency and authority, and it preserves the ambiguity that offers openings, however brief, to social subjects.

Finally, our theory moves beyond a notion of agency or authority that can be unproblematically invoked to explain the success or failure of a rhetorical performance. Like Burke, we suggest that the motive behind rhetorical events is a shifting relationship between constraints and resources. We think this framework can help rhetorical critics better understand the dynamics behind rhetorical events within the postmodern world. In some practical sense, too, it might help rhetors better gauge the opportunities for efficacious action and better position themselves in the relational practices that configure the conditions for action in the world. The kind of overlap and slippage between agency and authority we have suggested might help rhetors both understand the way regulative forces shape the terrain of social space and how this interplay opens possibilities in the grid of regulation. But beyond that, we anticipate that the application of this framework may well reveal yet other relationships between agency and authority.

We hope that the ambiguity that marks the overlap between agentive and authoritative spaces might also help us consider other more radical writing practices in other, less regulated cultural sites. As one of us works with zines (self published and counterculture magazines), we notice a similar interplay between agency and authority. Zines are illegitimate and nonstandard productions—irreverent, parodic, utopean, and imaginative. In some sense they are performances of difference that try to make a difference; that is, change subjectivity and representational practices and replace exclusionary and oppressive discursive practices. Such radical rhetorical performances constitute a *third space* that offers insight into the double or multiple-voiced discourses that characterize third space subjectivities (see Anzaldúa, 1987; Gates, 1998; DuBois 1986; Bakhtin, 1981). Zines and third-space subjectivity are far afield from the standard sites of analysis and research in professional communication, at least in English studies. But they enact the dynamics of discourse and social change on a continuum with more regulated institutional discourse. Radical rhetorical events like zines offer what Brownwyn Davies (1993) considers "disruptions [that can see] the possibility of breaking down old oppressive structures and of locating and experiencing [them] differently, of moving outside the fixed structures" (p. 39). If we think through the possible relations of agency and authority in all these sites, we might better understand institutional discourse as a genuinely cultural practice. Institutional discourse and zines are both constituted by the complex political and rhetorical relations of agency and authority we have begun to trace. If we can see similarities in the rhetorical dynamics of these kinds of sites, we might begin to think of the third space of imagination and ambiguity as a resource for subjects working within everyday institutional discourses.

REFERENCES

Althusser, L. (1971). *Lenin and philosophy and other essays.* (B. Brewster, Trans.) London: New Left Books.

Anzaldúa, G. (1987). *Borderlands, la frontera: The new mestiza.* San Francisco, CA: Aunt Lute Books.

Bakhtin, M. (1998). Discourse in the novel. In J. Rivkin & M. Ryan (Eds.), *Literary theory: An anthology* (pp. 32-44). Malden, MA: Blackwell Publishers.

Bloom, A. (1987). *The closing of the American mind.* New York: Simon and Schuster.

Bordo, S. (1998). Material girl. In J. Rivkin & M. Ryan (Eds.), *Literary theory: An anthology* (pp. 1099-1115). Malden, MA: Blackwell Publishers.

Bourdieu, P. (1977). *Outline of a theory of practice.* (R. Nice, Trans.). New York: Cambridge University Press.

Bourdieu, P. (1990). *The logic of practice.* Cambridge, UK: Polity Press.

Burke, K. (1969). *A grammar of motives.* Berkeley: University of California Press.

Butler, J. (1990). *Gender trouble: Feminism and the subversion of identity.* New York: Routledge.

Campbell, K. K. (2003, September). *Rhetorical agency.* Paper presented at the conference of the Association of Rhetoric Societies, Evanston, IL.

Carter, M. (1988). *Stasis* and *kairos*: Principles of social construction in classical rhetoric. *Rhetoric Review, 7*(1), 97-113.

Clifford, J. (1991). The subject in discourse. In P. Harkin & J. Schlib (Eds.), *Contending with words* (pp. 38-51). New York: MLA.

Condit, C. (1997). In praise of eloquent diversity: Gender and rhetoric as public persuasion. *Women's Studies in Communication, 20,* 91-116.

Condit, C. (1999). The character of "history" in rhetoric and cultural studies: Recoding genetics. In T. Rosteck (Ed.), *At the intersection: Cultural studies and rhetorical studies* (pp. 168-185). New York: The Guilford Press.

Davies, B. (1993). *Shards of glass: Children reading and writing beyond gendered identities.* Cresskill, NJ: Hampton Press.

de Certeau, M. (1984). *The practice of everyday life.* Berkeley: University of California Press.

D'Souza, D. (1991). *Illiberal education: The politics of race and sex on campus.* New York: Free Press.

DuBois, W. E. B. (1998). *The souls of black folk.* In J. Rivkin & M. Ryan (Eds.), *Literary theory: An anthology* (pp. 868-872). Malden, MA: Blackwell Publishers.

Ede, L., Glenn, C., & Lunsford, A. (1995). Border crossings: Intersections of rhetoric and feminism. *Rhetorica, 13,* 401-441.

Faber, B. (2002). *Community action and organizational change: Image, narrative, identity.* Carbondale: Southern Illinois University Press.

Foucault, M. (1972). *The archaeology of knowledge and the discourse on language.* New York: Pantheon Books.

Foucault, M. (1982). The subject and power. *Critical Inquiry, 8,* 777-795.

Gates, H. L. (1998). The blackness of blackness: A critique on the sign and the signifying monkey. In J. Rivkin & M. Ryan (Eds.), *Literary theory: An anthology* (pp. 903-922). Malden, MA: Blackwell Publishers.

Giddens, A. (1984). *The constitution of society.* Berkeley: University of California Press.

Goankar, D. (1997). The idea of rhetoric in the rhetoric of science. In A. Gross & W. Keith (Eds.), *Rhetorical hermeneutics: Invention and interpretation in the age of science* (pp. 25-88). Albany: State University of New York Press.

Grossberg, L. (1992). *We gotta get out of this place: Popular conservatism and post modern culture.* New York: Routledge.

Habermas, J. (1970). Toward a theory of communicative competence. In H. Dreitzel (Ed.), *Recent Sociology No. 2.* (pp. 114-148). New York: Macmillan.

Hegde, R. S. (1998). A view from elsewhere: Locating difference and the politics of representation from a transnational feminist perspective. *Communication Theory, 8,* 271-297.

Henry, J. (2000). *Writing workplace cultures: An archaeology of professional writing.* Carbondale: Southern Illinois University Press.

Herndl, C., & Bauer, D. (2003). Speaking matters: Liberation theology, rhetorical performance, and social action. *College, Composition and Communication, 54,* 558-585.

Laclau, E., & Mouffe, C. (1985). *Hegemony and socialist strategy: Towards a radical democratic politics.* London: Verso.

Leach, E. R. (1962). On certain unconsidered aspects of double descent systems. *Man, 62,* 130-134.

Longo, B. (2000). *Spurious coin: A history of science, management, and technical writing.* Albany: State University of New York Press.

Mailloux, S. (2003). *The role of rhetoric in the academy and beyond.* Paper presented at the conference of the Association of Rhetoric Societies, Evanston, Illinois.

Messer-Davidow, E. (2002). *Disciplining feminism: From social activism to academic discourse.* Durham, NC: Duke University Press.

Morris, M. (1990). Banality in cultural studies. In P. Mellencamp (Ed.), *Logics of television: Essays in cultural criticism* (pp. 14-43). Madison: University of Wisconsin Press.

Pérez, E. (1998). Irigaray's female symbolic in the making of chicana sitios y lenguas (sites and discourses). In C. Trujillo (Ed.), *Living chicana theory* (pp. 87-101). Berkeley, CA: Third Woman Press.

Pérez, E. (1999). *The decolonial imaginary: Writing chicanas into history.* Bloomington: Indiana University Press.

Poulakos, J. (1983). Toward a sophistic definition of rhetoric. *Philosophy and Rhetoric, 16,* 35-48.

Sandoval, C. (2000). *Methodology of the oppressed: Theory out of bounds.* Minneapolis: University of Minnesota Press.

Smith, P. (1988). *Discerning the subject.* Minneapolis: University of Minnesota Press.

Sommers, C. H. (1994). *Who stole feminism?: How women have betrayed women.* New York: Simon & Schuster.

Spivak, G. (1988). Can the subaltern speak? In C. Nelson & L. Grossberg (Eds.), *Marxism and the interpretation of culture* (pp. 271-313). Urbana: University of Illinois Press.

Spivak, G. (1999). *A critique of postcolonial reason: Toward a history of the vanishing present.* Cambridge, MA: Harvard University Press.

Sullivan, P. A., & Porter, J. E. (1993). Remapping curricular geography: Professional writing in/and English. *Journal of Business and Technical Communication, 7,* 389-422.

White, H. (1990). *Tropics of discourse: Essays in cultural criticism.* Baltimore, MD: Johns Hopkins University Press.

Williams, R. (1977). *Marxism and literature.* New York: Oxford University Press.

CHAPTER 8

Rhetoric of Empowerment: Genre, Activity, and the Distribution of Capital

Dave Clark

In this chapter, I examine the empowerment narratives that frequently accompany the emergence of new technologies and management philosophies. These narratives have most recently shaped our reception of knowledge work, which is assumed to be inherently empowering because it provides greater access to critical organizational information and decision-making authority. I rely on activity and genre theories as well as an application of Bourdieu's notion of capital in suggesting not only that knowledge workers are not inherently empowered—as others have argued—but that worker empowerment, a fairly new management concept, is regulated by information access, but also by position, education, profession, and the solidification and dissolution of organizational networks. In doing so, I hope to contribute an understanding of not only the means by which empowerment is defined, created, and accessed, but of how a cultural, rhetorical approach to qualitative research can illuminate the regulatory functions of organizational discourse.

[E]mployees are armed with a handheld Telxon gun that scans the product bar code on each item and gives sales information on the display readout. Sales for the day, week, and five weeks back give the associate, as Wal-Mart staffers are known, the sales history of the item. There's also an instant readout of the number of items on the shelf, in stock, in transit, and on order. "That's empowerment," says Bill Woodward, vice president and chief administrative officer for Wal-Mart International. "It is the link to life at Wal-Mart. If you expect people to do things and you give them the tools, they'll do it even better than you expected." (Tapscott, 1995, p. 153)

Myopic empowerment narratives like the one above are fun and easy targets. Anyone without an enormous stake in the public perception of Wal-Mart would be hard-pressed to see empowerment in this story for a low-salary, no-benefits employee. But the claim of empowerment itself is not surprising. Many popularists, analysts, and scholars of contemporary information-age, post-Fordist, knowledge-based enterprises have argued that while Taylorist organizations historically hired nonmanagerial employees only from the neck down, new post-bureaucratic organizations have redistributed knowledge and information, granting employees new rewards and autonomy by giving them access to the means of decision making. These employees, suggest some, have access to the means of production beyond "what Marx could have dreamed of" (Tapscott, 1995, p. 67). In these new organizations, even shop-floor employees are empowered with decision-making authority and a greater sense of ownership; information-age employees are hired from the neck up as well as from the neck down.

Gee, Hull, and Lankshear (1996), among others, have argued that this empowerment is simply a more efficient Taylorism that simultaneously activates employees' brains and convinces them that they have been empowered, improving the ability of an employee to quickly perform management's tasks in management's fashion without any significant gain in personal rewards or autonomy (pp. 30-31). Empowerment is, in their argument, more accurately described as a management fantasy about increased efficiency. Gee et al.'s articulation of knowledge work strikes me as accurate, and many others have also argued (cf. Aronowitz & DiFazio, 1994; Greenbaum, 1998; Hull, 1999; Rifkin, 1995) that many classes of knowledge workers (e.g., support-line operators, data-entry clerks, assembly-line workers, entry-level computer programmers) have not been empowered in any meaningful, material way through new and improved access to information. At the same time, some workers have had their jobs irrevocably changed by technology in ways workers themselves claim as empowering; they are shuffling information all day, and despite the widely reported (cf. Hayes, 1995; Tenner, 1997) negative effects of technology on bodies and organizations, at least they are not digging ditches.

In other words, contemporary empowerment narratives are insufficiently nuanced to address the complexity of empowerment. At the same time, responses to the narrative have failed to adequately embrace the best of what we know about how information—texts, in a broad sense—is connected to more complex and localized distributions and applications of power. In this chapter, I offer notes toward an updated understanding of information-based empowerment that rejects the unsophisticated, one-dimensional understanding of power proposed by contemporary empowerment narratives in favor of a cultural approach that attempts to assess the complex structures that govern empowerment. To build that approach, I borrow from activity theory and genre theory as well as from Pierre Bourdieu's notion of capital in order to question not only whether workers are empowered, but also what regulates their access to that empowerment.

I begin by examining recent empowerment narratives, suggesting they are part of a long rhetorical tradition. Next, I elaborate the methods and theories that I hope will lead to a more nuanced understanding of power in information-based organizations. Finally, I apply those theories (specifically a synthesis of genre and activity theories) with Pierre Bourdieu's sense of capital to the task of analyzing the power arrangements at my research site. I narrow my research question down to this: How empowered—and in what ways empowered—are knowledge workers in a contemporary, Internet-savvy organization? Following my analysis, I argue that worker empowerment is, of course, far more complex than simple access to information, and it is also more complex than imagined by critics of empowerment. Workers are regulated, that is linked to implications of their work and decision-making authority in their work, in ways that structure and are structured by the mutual constitution of actors, genres, and capital, and by the solidification and dissolution of organizational networks.

EMPOWERMENT NARRATIVES

There is a long history of empowerment narratives associated with the birth of new technologies. Carolyn Marvin suggests in *When Old Technologies Were New: Thinking about Electric Communication in the Late Nineteenth Century* (1990) that Victorian views of the telephone and the electric light were similar to our perceptions of the personal computer and the Internet; and Tom Standage's *The Victorian Internet: The Remarkable Story of the Telegraph and the Nineteenth Century's Online Pioneers* (1999) similarly points to parallels between telegraph and Internet development; the telegraph, too, was imagined as a device that would democratize communication and empower heretofore powerless individuals to act independently of mass-imposed structures that had dominated their lives. Narratives of empowerment and democratization have been associated with nearly every new technology since the printing press, including the telephone, radio, and television, all of which were to provide unprecedented individual power and access. It is safe to say that the long-term implications of these technologies have been more complicated than initially imagined.

Empowerment narratives continue to pop up whenever new technologies are in the process of being sold in our culture, perhaps because the narrative is uniquely appealing to American eyes and ears. Consider the parallel of these empowerment stories with many mainstream American films that tend to portray ragtag bands of clichéd and stock individuals triumphing against all odds, including hegemonic monoliths; this is the basic narrative of films from the 2003 release *Seabiscuit* (in which variously damaged individuals and a too-small, troubled horse with heart succeed in toppling War Admiral's blue-blood empire) to *The Matrix* (in which a ragtag band fights the machines that, it turns out, rule reality) to pea-brained '80s action films like *Rambo* and *Missing in Action*.

In the case of the Internet, the empowerment narrative was particularly pervasive. In the late '90s, scores of books, magazines, and scholarly articles argued for its inherent empowering effects. Drawing from earlier information-age texts like Peters' (1994) *Liberation Management: Necessary Disorganization for the Nanosecond Nineties* and Senge's (1990) *The Fifth Discipline: The Art and Practice of the Learning Organization*, these texts argued that not only did the Internet make knowledge work both possible and inevitable, it created new empowerment possibilities across the board. The developing new economy of the information/Internet age was to be a postindustrial economy in which mass production, mass consumption, and the mass media are ineffective. Networked media empower consumers to make customized choices rather than brainwash them into buying standardized products. Organizational hierarchies, and in turn big government and big business, crumble before the relentless pressures of the empowered customers of the free market. The key word throughout the story is empower. Home life, work life, and social life are all improved because individuals are empowered to manipulate knowledge and make free choices rather than be subject to the control of bureaucracies, government/corporate influence, or to consumer inconveniences. Simply by giving our workforce computers, computer expertise, and the Internet, we create empowered workers who are more efficient, productive, and better consumers of the products of our industries.

Gee et al. (1997) devote Chapter 2 of their *The New Work Order: Behind the Language of the New Capitalism* to a critical retelling of this narrative. They characterize it as the fast capitalist story of the new economy, a utopian story that is told and retold in dozens of sources, including "texts produced mainly by business managers and consultants [who] seek to attend as textual midwives at the birth of the new work order" (p. 24). The general uniformity of vision in these texts is widely noted and described by James Boyle (1997) as "crude but passionate libertarianism" (pp. 6-7), and subject to satire in Dilbert comic strips and by online columnists. Indeed, the most popular and influential of the late '90s popular-technology texts (including Gates' *The Road Ahead* (1995), Negroponte's *Being Digital* (1995), Dertouzos's *What Will Be: How the New World of Information Will Change Our Lives* (1997), Dyson's *Release 2.1: A Design for Living in the Digital Age* (1998), Tapscott's *The Digital Economy: Promise and Peril in the Age of Networked Intelligence* (1995), virtually every issue of *Wired* magazine, numerous speeches by Newt Gingrich and Al Gore, and prevalent advertisements like AT&T's "You Will" ads and Xerox's "Share the Knowledge" ads) all tell a similar "techno-utopian" (Doheny-Farina, 1996, p. 14) story.

The *dot com* bust of the late '90s effectively killed the authorship and demand for many of these utopian texts, and bookstore Internet bookshelves are now weighed down by "Dummies" books and technical manuals, a fact that empha-sizes the extent to which the technology itself has been successfully integrated into our culture. But the empowerment narrative lives on, moved just a few shelves over to sections devoted to Linux and other open-source technologies (featured in IBM's utopian 2004 advertising campaign), and to particular network-based uses of Internet technologies like content- and knowledge-management systems. Texts like Raymond's (2001) *Cathedral and the Bazaar: Musings on Linux and Open Source by an Accidental Revolutionary* and Pavlicek's (2000) *Embracing Insanity: Open Source Software Development* emphasize the power of small groups of individuals to overcome and write better, faster code than the hegemonic forces of corporate software creators (particularly Microsoft) through organized, market-driven action. In the books themselves, the democratizing rhetoric is both obvious and pervasive. Meanwhile, texts like Drucker, Garvin, Dorothy, Straus, & Seely Brown's (1998) article in the *Harvard Business Review of Knowledge Management* and Davenport and Prusak's (1998) *Working Knowledge: How Organizations Manage What They Know* emphasize the empowerment that can come from knowledge-management initiatives that use corporate restructuring and Internet-based technologies to deliver on-demand information to knowledge workers.

So the narrative marches on, finding subtechnologies and new technologies on which to focus, and we are still without a nuanced means of discussing it. In what follows, I do not claim to have refuted the enormously complicated business of empowerment, which is a slippery concept. But in the next section, I define my cultural approach to the problem and describe the methodologies that I believe can at least lead to some more workable, problematized definitions we can employ in assessing the relative empowerment of knowledge workers.

APPROACH: THE CULTURAL TURN

The argument surrounding the "cultural turn" can be explored with a few Googles, which yield economists who (to me, wrongly) equate the cultural turn with postmodernism, sociologists who see it as a move beyond foundational understandings of the social, and even texts from historians. In *Beyond the Cultural Turn* (Bonnell & Hunt, 1999), the editor's introduction suggests that for historians the social was an epistemological phase of the '60s and '70s in which scholars relied on "research paradigms that proposed to organize the study of society on the model of the natural sciences" (p. 1). These paradigms are based on what the authors see as a commonsensical, unstated view of the social that took social categories as givens and attempted enormous scientific studies of those categories. The cultural turn, which they see as occurring in the early '80s in the social sciences, was a move to the linguistic, microfocused models of Geertz, Foucault, and Bourdieu.

In my field of rhetoric and professional communication, the links to cultural theory had their origins in responses during the 1990s to Kenneth Bruffee's

(1984) discussion of social construction, most notably in the release of Blyler and Thralls' (1993) edited collection *Professional Communication: The Social Perspective;* their volume raises ideological responses that critique Bruffee and others for their relatively neutral and naïve approaches to power and agency (p. 14). More recently, Bernadette Longo (1998) picks up the thread by indicating what she sees as the possibilities of cultural studies to eliminate what is lacking within social construction: "Applied to technical writing, [a Foucauldian cultural approach] can illuminate how struggles for knowledge legitimation taking place within technical writing practices are influenced by institutional, political, economic, and/or social relationships, pressures, and tensions within cultural contexts that transcend any one affiliated group" (p. 62).

In other words, our focus on the social has been driven by our attempts to analyze the ways that groups and individuals use language; that is, the ways language works in given social contexts. There has been little room in such discussions, Longo suggests, for agency and power (p. 54). As Winsor argues, such analyses are "incomplete because [they] ignore the circumstances in which much knowledge work is done, that is, in for-profit, hierarchical corporations. . . . While Hutchins's (1993, 1996) examination of distributed cognition calls our attention to the fact that knowledge is often communally held and thus depends on communication, we must remember that systems of distributed cognition are not always collaborative, egalitarian, and harmonious" (2003, p. 7).

Weaknesses in a purely social approach include a lack of emphasis on power and ideology, the epistemological failure of social construction to articulate connections to the nonlinguistic material realm, and the typical, if unnecessary, disconnect of social analyses from study of larger cultural forces. These weaknesses have led professional communication scholars to seek approaches that provide broader cultural analyses. For example, Charles Bazerman (2003) suggests that "rhetoric has a limited model of the functions of writing," and that "[o]ur understanding of writing can be greatly enriched if we step beyond traditional rhetorical approaches. . . [using new] tools of research and theory" (p. 452) drawn from the social sciences. Many social constructionists might not see themselves in Bazerman's description of rhetorical researchers "reliant on a discipline directed towards high stakes agonistic public performances (primarily spoken) having to do with policy choices, the adjudication of disputes, and the forging of communal values" (p. 25); but his point about expanding our research tools and possibilities is nevertheless a good one.

Certainly an investigation of workplace-empowerment narratives could benefit from a social-science approach that was designed to deliver nuanced discussions of power. After all, empowerment is difficult to pin down; what counts as empowerment varies widely from industry to industry and, most of all, from individual to individual. I think it is reasonable to expect that an assessment of empowerment should include, at minimum, a broad sense of the reward system afforded by the job, including personal sense of autonomy, pay, benefits, the ability to choose tasks and structure one's day, and the value placed on work by colleagues. The variability of empowerment has been noted in studies critical of the inherent empowerment of the new knowledge worker (cf. Gee et al., 1997; Rifkin, 1995). But to conduct an analysis that focuses on the variability of empowerment, I need a theory well-designed for the purpose: Bourdieu's concept of capital, which in his terms suggests not only monetary capital but also social, symbolic, and cultural capital (*Distinction*).

Bourdieu (1987) suggests that individuals are inculcated from birth with a "habitus," a "system of durable, transposable dispositions . . . principles which generate and organize practices and representations" (p. 14), a set of deeply internalized precepts. Bourdieu further suggests that action is shaped by the interaction of an individual's habitus and his or her "field," the social context in which she or he operates, defined as "a structured space of positions" in which the positions and their interrelations are determined by the distribution of different kinds of capital (p. 14). The interaction between habitus and field is complex; a back and forth negotiation exists between the context and the individual. Moments of agency also exist, but they are structured by what is allowed. Entering a field requires accepting the underlying suppositions of that field; at my case-study organization that means, in part, working within the confines of the agency and power limits of the organization. Bourdieu sees power, then, as a product not simply of hierarchical structures, but of the intersection of a habitus and a field, an intersection that structures the types of capital held by individuals and the rates of exchange allowed for their capital. For example, a software developer can exchange her education (cultural capital) and the value placed on her work (symbolic capital) for a high salary (monetary capital) or for freedom of movement and task selection (social capital). But a marketing intern, who is seen as possessing less cultural and symbolic capital, usually receives less monetary and social capital in exchange for her efforts.

These concepts, directed to a workplace study, can indicate with some nuance the value placed upon an employee's work and the autonomy he or she has within the job. For example, as I discuss in more detail in my case study below, if one examines technical writers and marketers from the perspective of capital, their work—"knowledge work" and therefore "empowered" work by any pundit account—provides limited freedom for choosing and structuring tasks, forces them to address and embrace the uncomfortable ideologies of the products and services they supply, and offers relatively poor rewards. The company's technical workers, by contrast, had more freedom, had the luxury of isolating their work from the politics of their products, and had access to far more resources.

So Bourdieu can help us with articulating types and amounts of rewards. As a means of analyzing the activities in contexts in which capital is accumulated, I use a combination of postsocial construction theories—genre and activity—to give a more formal structure to my analysis of SecureCom's culture and to make use of theories that Russell (1997b) argues are well-suited to analyses of power: "To understand power in modern social practices, one must follow the genres, written and otherwise. Power appears in specific, locatable occasions of mediated action and is created in the network of many localized instances. Power is analyzable in terms of the dialectical contradictions in activity systems, manifest in specific tools-in-use (including written genres) that people marshal when they are at cross-purposes" (p. 524).

Genre theory is then uniquely well suited for analyses of the networks of linked texts in an organization such as my case-study site. And Bazerman (2003) suggests that activity theory can help enrich perspectives otherwise based solely on genre (p. 37). Activity theory, which derives from the work in the '30s of Russian social psychologists Vygotsky and Leont'ev, has been adapted and used in professional communication, through the work of Cole and Engeström (1993). Activity theorists argue that an "activity system," the basic unit of analysis, is "any ongoing, object-directed, historically conditioned, dialectically structured, tool-mediated human interaction," for example, "a family, a religious organization, an advocacy group, a political movement" (Russell, 1997b, p. 510). Groups and individuals are analyzed with a triangular approach that emphasizes the multidirectional interconnections among subjects (the individual, dyad, or group), the mediational means or tools they use to take action (machines, writing, speaking, gesture), and the *object* or problem space on which the subject acts (Russell, 1997b, p. 510). Multiple activity systems overlap and interconnect, just as occurred with discourse communities in social construction. Other, related theorists have noted that subjects need not be human (cf. Latour, 1991; Myers, 1996), that machines, texts, and other objects can also act. In addition, activity theories can provide us with a broader cultural understanding of an activity system. As Flower (2003) suggests, "In Engeström's (1993) powerful model [activity analysis] includes not only the actors, the object of action, and the community which shares those objects, but also divisions of labor or power, the rules and conventions, and the material or symbolic tools that mediate the activity" (p. 242).

My use of these theories is not unique; many scholars recently have been combining activity and genre theories, and more. For example, Bourdieu's perspective has become common in professional communication scholarship, particularly his concept of capital. But I do suggest here that this combination of theories is well suited to the task at hand, and in what follows I use a combination of Bourdieu's theory of capital, genre theory, and activity theory in an attempt to assess the empowerment of workers in an Internet startup.

Research Setting and Methods

SecureCom,[1] the site of my investigation of empowerment, is a small company located in a research park at a Midwestern university. I spent 15 months between January 1999 and March 2000 gathering data. At the time of my study, the company's sole product was ProBlocker, a hardware/software bundle designed to allow schools, libraries, and businesses to protect their networks and block access to "inappropriate, objectionable, or unproductive" (SecureCom marketing materials) sites on the Internet. The four cofounders were the only employees: Dan, a professor of computer engineering at the university; his wife, Gina, who worked full time as a tech-support engineer; Susan, who worked full time as a developer; and her husband, Glen, who built the hardware. Fortunately for SecureCom, sales and buzz about the product were strong enough that they were able to interest investors, and in the course of my study they acquired venture capital, moved into new office space, and hired 12 more employees. Their success, growth, and transformation were a stroke of luck in my research, allowing me to observe the construction, revision, and solidification and codification of their material and linguistic structures as new hires, new work. New texts became necessary and were carefully structured and enlisted both to isolate uncomfortable workers from the political implications of their work and meet expectations of the corporate genre. As several participants put it, it was "the way things are done."

During the first six months of my study, I observed the individual work sites of the four owner/employees. I made 2–3 visits a week, averaging about 10 hours of observation a week during the first few months of my study. I took roughly 100 pages of detailed field notes on the environments, work practices, and the infrequent telephone calls. I also conducted daily ad hoc interviews about the day's work and collected documents and e-mail messages, primarily on the basis of what my participants found interesting and useful enough to share. As is common in ethnographic research (cf. Agar, 1980) my own research questions had not yet fully developed. I therefore allowed my participants to shape my research with their contributions. They knew I was studying communication—the evolving nature of my ethnographic work made a more detailed assessment difficult, although I informed them as my research progressed—and they would frequently volunteer work and documents they felt I would find interesting.

Later in my study, as the company moved into new office space and gradually hired 12 new employees (including me as a contract technical writer), my research environment changed significantly. I observed and took field notes on all of the new employees as well as on the changing and evolving work, procedures, and politics of the organization. Finally, as I neared the end of my arbitrarily determined 15-month data-gathering period, I conducted half-hour

[1] The company name, product name, and names of all individuals are pseudonyms.

interviews with all 15 of the participants; as my research questions had now narrowed to focus on the construction of what I describe below as the technical/ nontechnical boundary, I was able to specifically target my interviews to gather their interpretations of the technical/nontechnical questions I was attempting to address in my research.

Also critical here is my own participation. Ethnographic researchers inevitably participate in the activities they articulate as well as in the articulation of those activities. My participation, however, went a step beyond, as I was hired by the company as a technical writer only three months into my time there as a researcher. Being hired made my access to data easier, but it also created ethical complications. First, my status as a consultant, then as an employee made me fully complicit in the activities of the organization, thus complicating easy conclusions but also making negative critique substantially more difficult. I thought mounting such a critique might be necessary; when I began my research, I knew that the company, as a censor of content in educational institutions, engaged in practices that in many ways I disagree with. From my own ethical perspective, then, there could well have been negative stories at and about SecureCom that needed to be told, depending on the direction my research took. Failing to tell these stories would mean risking uncritically "reproducing the social structures, ideologies, and subjectivities" (Herndl, 1993, p. 353) of SecureCom. Further, my working at SecureCom problematized my relationships with my participant/coworkers, building their trust in me and making them more likely to share with me things they might not otherwise have shared. My position in the company, then, made my position as a researcher less ambiguous and made the potential political threat of my research less threatening in the eyes of my participants.

The Technical/Nontechnical Split

SecureCom's activity system involves a staff of people leveraging a wide-ranging tool set (capital, computer equipment, physical space, human networks) towards the object of producing Internet filters and turning a profit. That system, of course, could also be described as the sum of dozens of smaller activity systems directed at smaller, discrete tasks that add up (or do not) to the end goals of the organization. For example, SecureCom's programmers, working on just the user interface or on just the machine-level blocking code, use their tools—writing, expertise, software libraries, computer equipment, collaboration—to produce management-designated pieces of products. But as Russell (1997b) points out in his analysis of the activity systems of cell biology, such systems are always embedded in and linked to other systems. In the case of cell biology, these are linked to the public, to government research agencies, to advocacy groups, and to drug companies. SecureCom is implicated in a number of influential activity systems, including standard instantiations of the genre of corporation, the software industry, the filtering industry, culturally conservative beliefs and platforms, the educational system that produces employees, legislation that governs filtering, and the politics and policies of their target-market schools and libraries. They are also situated by their location within a university research park, an organization that by its charter emphasizes the application of basic technical research to the production of products and profit.

I suggest here that in many ways the story of empowerment at SecureCom can be usefully paralleled to the organization's adoption of a linguistic and material activity structure typical to software development organizations that I term the technical/nontechnical split. The split is a commonsense separation of technical work from other work performed in the organization. The separation, I suggest, structures access to other organizational genres and activity systems in ways that govern the acquisition of capital by SecureCom employees. Many (even most) U.S. organizations have highly developed technical/nontechnical splits that have enormous impacts on day-to-day life and work within those organizations. The so-called new economy of the information age has stabilized and accelerated the formation of those splits. New technical divisions have been formed both in educational institutions and in corporations that exist to supply the technological infrastructure necessary for this knowledge work. As in the past, technical schools train students in trades like truck repair, system administration, and computer programming. Increasingly, universities offer similar programs in addition to their always-present production of self-identified technical people like engineers and programmers. The corporate world creates, mirrors, and thereby solidifies these structures in that virtually all corporations now have a technical side; that is, an information technology department responsible for upkeep of the organization's computer systems. Further, a service industry of hundreds of companies has organized solely to provide consumers and other businesses with now-essential computers and software.

Even as I describe the split, I am trying not to assume its existence; a technical/nontechnical split, while common, is not natural or a given. That is, people or objects are not inherently or exclusively technical or nontechnical except as they are defined through human articulation. In fact, a significant line of scholarship argues that technologies and technologists are not exclusively "technological," but are in fact technosocial, constructed, and given status and meaning as technical only within social contexts. When one considers this argument—with which I agree—it is difficult to maintain the common, day-to-day separation between technical people and objects as well as nontechnical people and objects that is my focus here; even those not familiar with this argument can be easily persuaded to see the problems with this dichotomy. For example, after I noticed the strong impact of technical thinking on the structure at SecureCom, in the interviews I conducted in my SecureCom study, I asked participants to elaborate on their concept of what technical meant and who they considered to be a technical person. The software developers I interviewed, confident in their own positions as technical people, began with a definition that essentially described themselves and created a continuum of "technicalness," tying technical to computer and mathematical prowess and, as in Ellen Ullman's (1997) *Close to the Machine: Technophilia and its Discontents,* to the amount of interaction an individual had with the guts of machines themselves. For example, they rated a developer who works with machine code as more technical than one who develops the user interface, and in turn they rated interface developers as more technical than users or a help desk operator, both of whom must use and understand the machine but do not program it. The developers saw technical writers and technical-support personnel as boundary crossers, neither completely technical nor completely nontechnical, and indeed, there is much blurring of the boundary, particularly as it is initially being formed. However, they were hesitant to see people who worked farther from machines (in this case, desktop computers) as technical; the marketing director, after all, was not very proficient with her e-mail software, and the VP could not always keep up in technical conversation.

When pressed even slightly, however, participants' definitions broadened significantly; the SecureCom accountant was technical because he worked with numbers; as the firm's technical writer I was technical both because I worked with technology myself and because I possessed a technical understanding of the way language operates, and even the marketing director was technical because she possessed carefully honed knowledge and skills from a specialized discipline. In other words, when challenged, their stock notion of technical was easily broadened to the point that it barely held on to their original meanings and certainly allowed them none of the initial prestige they seemed to draw from the term. Technical quickly moved from "connected to the machine" to "having specialized skills." Thus, even for people who have a significant stake in the hierarchy implied in their use of the technical/nontechnical boundary, the boundary is easily permeable, and finally, few people and objects are not seen as both technical and nontechnical. Nonetheless, the technical/nontechnical divide is pervasive; my participants were surprised by how easily holes could be poked in their definition of "technical," but they were equally surprised that I had even questioned their definition. The conventional definition of technical had been naturalized to the point that it was part of everyday conversation and, therefore, did not need to be defined.

The term technical is used every day by SecureCom workers—and workers everywhere—to differentiate their work, their machines, their understandings and those of others. Nontechnical, by contrast, is a less frequently used term that in most cases means, simply, the lack of technical. Nontechnical, too, is a highly contextualized term, although we seldom think of it that way. An engineer who works on generating power from hydraulic dams would doubtless be considered technical in that specific sense, but, lacking specific skills that would enable her to get close to the machine, could have only a nontechnical role at SecureCom. Nontechnical workers are almost never referred to as nontechnical; their work simply lacks the prestigious technical tag that would (as we will see) cause it to be more highly valued within the organization.

In what follows, my definitions of technical and nontechnical rely on my participants' conventional uses of the terms. Why do these conventional definitions matter? They matter because even though the definitions are arbitrary and easily challenged, as Bowker and Star (2000), Downey (1998), and others argue, they do significant work in organizations. At SecureCom, the technical/ nontechnical split worked in collaboration with writing to determine the way the company grew, both defining the way writing was used and itself being defined by company writing practice. Early in my interviews, the split was active for my participants, all of whom were technical people, and this split impacted the ways they decided to develop the company in that their own understandings of technical vs. nontechnical drove the ways they divided labor, space, and writing tasks. Those divisions in turn created and perpetuated the split on a company level, as the human actors enlisted writing as a technical tool, as a means to help them create the desired network for building and selling their technology. Over time and through the development of the company and the split, writing was also enlisted to serve a regulatory function and to further create and perpetuate the split as it was technologized to regulate both official and unofficial social arrangements.

Like SecureCom, many companies have a mature technical/nontechnical split that structures their work and reward systems. In software firms, computer-programming departments are physically isolated from others in the company to avoid distractions from the less technical work of other workers; at one local company, developers were physically separated by a door that required key-card access; at another, nondevelopers were allowed in the development area only during morning zoo hours at the beginning of the day, after which they were required to leave the area (suggesting that the presence of nontechnical staff was seen as a distraction). The split also structures work in smaller but equally tangible ways. At SecureCom, for example, development's technical interns with technical experience but without college degrees were routinely hired at $10 per hour; technical-writing and marketing interns were hired at as little as $7 per hour and were obliged to fight for increases after obtaining degrees that, according to company policies, should entitle them to raises. The justification was the self-fulfilling prophecy of market forces based on the split: the notion that technical jobs should and would be more highly rewarded because of the availability of personnel with such credentials (despite the relative ease of finding technical employees so close to a university of science and technology).

A missing part of this story, I suggest, is the value placed on certain kinds of work through use of the split; being assigned a technical genre of work that carried high symbolic capital, such as authoring machine code, guaranteed employees access to activity systems with greater capital rewards. Being assigned a less technical (but still technical) genre such as coding a graphic user interface

(GUI), offered lesser (but still significant) rewards along with the social capital, to successfully change activity systems. This is where Bourdieu (1987) becomes very useful. Bourdieu's conception of capital, which incorporates monetary, social, symbolic, and cultural capital, helps account for the value placed upon an employee's work and the freedom and autonomy an employee has within her or his job. I suggest that this capital distribution is worth studying, as its very structure argues against the implicit empowerment of knowledge work suggested by many economists, social scientists, and popular culture theorists. Above, I define empowerment as including a personal sense of autonomy, pay, benefits, the ability to choose tasks and structure one's day, and the value placed on work by colleagues. At SecureCom, technical writers and marketers—knowledge workers by all assessments—had limited freedom to choose their tasks, were bound by their job categories to address and embrace the ideologies of the problematic products and services they supplied, and were relatively poorly rewarded for their work. SecureCom's technical workers, by contrast, had more freedom, had the luxury of isolating their work from the politics of their products, and had access to far more resources.

Capital and the Split

Elsewhere (Clark, 2001), I more fully develop my account of the technical/ nontechnical split, the story of how the split was adopted/created/ imposed at SecureCom. The split existed for several reasons, including participants' sense of how things are customarily done in corporate settings; their sense of disciplinary boundaries and their own strengths and weaknesses; and the values they placed on different kinds of work, which became instantiated and reified through the split. I am not suggesting here that SecureCom's decisions were wrong and that the company would have been better off with human resource managers attempting to program. My point is simply to examine the way that the boundaries established by the split structured work and capital for SecureCom employees.

Some background is appropriate. The story of SecureCom's split is the story of the adoption of genres and activity systems. Early in the company's history, certain genres and activities, such as the creation of hardware and software code were viewed as technical tasks that should be the work of technically trained people. Other tasks, such as corporate correspondence and record keeping, were necessary but undesirable nontechnical tasks; they were necessary because my participants viewed them as part of the way things are done in the computer industry. Interestingly, the bulk of the transformation was about the redistribution of genres of work thought to be outside the purview or the abilities of the company principals. The principals rid themselves of business record keeping, much of their informal correspondence, and the Web site, leaving themselves only with internal e-mail messages, the technical documents (e.g., white-board sketches, the occasional white paper, and internal documentation) necessary to organize the development team, and the actual code that they wrote in developing the product.

The result of these changes was a redistribution of capital rewards, which previously had been distributed fairly equally among the four principals. The new titles and hierarchies can allow us to make some judgments about capital distribution. It is not hard to guess, for example, that the chief operating officer (COO) has significantly more cultural, material, symbolic, and social capital within the organization than the technical writing or marketing interns, the lowest paid, least valued, and least empowered employees within the organization. It would be, however, a mistake to assume, for example, that the director of marketing possesses more material and symbolic capital in the organization than I did as the technical writer simply by virtue of her higher placement in the organizational chart. In fact, my salary was equivalent, and my work was considered moderately important. It would be a mistake, as well, to assume that the development intern's social, symbolic, and material capital matched those of the marketing intern, despite her degree status and her social skills. He was granted significantly more of each. In what follows, I use the technical/social split to demonstrate the relative empowerment of SecureCom's workers.

Empowerment and Technical Workers

In building SecureCom into an instantiation of the technical/nontechnical split, the principals made decisions about the division of labor; as technical people, they were accustomed to directing their activity systems at machines. The restructuring of the organization, it turned out, was about creating a complex division of labor that allowed them to reassert their technicalness and eliminate tasks for which they felt ill-suited, indifferent, or hostile. In the process, the organization developed multiple new activity systems; whereas before the addition of new staff, principals used all their technical and nontechnical tools to strive toward multiple objects, now new staff members had targeted objects to pursue: customers, money, machines. Among the early activities of the new COO was the creation of policies, procedures, and standardized forms to govern activity. As the systems of genres expanded and became more and more specialized and codified, there was less and less slippage across job roles, and the technical/nontechnical split was instantiated.

All new technical employees were placed in the new technical room with the other developers. It was a space specifically designed with the production of their particular genres in mind. As with nontechnical employees, new technical employees were given individual cubicles, individual computers and phones, and focused, individualized, and specialized tasks designed to suit their particular talents. In addition, SecureCom installed a number of technical systems designed to lessen the amount of social interaction necessary for their work; for example, a source-control system managed the collaboration necessary for the development of large software projects. The basic assumption of technical work, then, was that technical workers would work with their heads down, fixed on the screen, interacting with others as little as possible. If it were not for organizational firewalls and the need for network access, the work could have been conducted from anywhere. Isolation was a theme of the design of SecureCom's new development process. Technical workers were physically isolated from the distractions offered by the constant interactions and phone calls of the sales and marketing staff, and cubicle walls were chosen that would minimize noise and distraction. Relative isolation, management assumed, would help efficiency and productivity by allowing the developers to focus on the isolated, decontextualized task at hand. One result was that when I asked technical workers about the politics of the filtering product, they often had not thought much about it. As Dan said,

> I guess it has ethical issues. I am not sure I've given that a lot of thought. I don't worry about the product in the sense that "I'm really making a difference in the world." I think it's just a product that's there for people who want to filter Internet access for whomever. If they want to do that, that's fine. I don't have a problem with it, but I wouldn't necessarily recommend it.

Dan's ambivalence was echoed by many of the technical workers, who never thought much about the impact the product had on the world; their day-to-day work, after all, was mostly about designing menus and fixing bugs, not "filtering" in a large sense. Waldo, like Dan, expressed ambivalence toward the ultimate filtering goals of the product, but he did not worry much about it.

> For my particular role, I'm not even thinking about the overall product, I'm thinking about little bits and pieces of it I'm making that work to the best of my ability. Sometimes when you step back and you do major testing, you do think about the picture as a whole, but it still doesn't matter to me because in the testing phase I'm making sure that my part works right.

So the work genres assigned to the technical workers did not, understood one way, allow them the luxury of thinking of the larger politics of what they worked on. Understood another way, it allowed them to ignore those larger politics as they chose and still participate in discussions of those politics and weigh in on marketing choices when they saw fit. On a daily basis, the goal of developers was to use their tools to write and improve code by working on concrete goals complete with correct answers (it worked or it did not). This work was seen as mission critical to the organization because it is (increasingly) possible to survive without a marketing department. As one intern, Hank, told me, "Marketing is unnecessary; we could put up a Web site and sell these things ourselves." Waldo added, "Without the meat, you've got nothing else." The principals and managers saw the end object of the firm as the generation of products and money. Thus, mission-critical tasks were the ones that received the biggest rewards. Put another way, the technically based activity system of the organization rewarded the production of technical genres with significant capital.

To be more specific, technical workers had enormous social capital and were able to participate or not (as they saw fit) in the politics of the organization's product, and they had considerable control over the boundaries and focus of their own work. For example, in addition to writing e-mail messages to each other and to developer help lines and chat rooms, the developers wrote two kinds of texts: computer code and technical notes. The notes included feature lists, schematics, flowcharts, and other reference and design documents that the technical people wrote together to describe the product(s) they were jointly creating. These texts, particularly the product feature lists, were frequently influenced by the desires and expectations of the nontechnical staff, but the developers had the last word on features that would be included, as they were imagined to be the only ones who could make rigorous assessments of the features' technical viability. The computer code was the guts of the product itself, and in its creation the developers received little to no external feedback, apart from the occasional GUI change.

The developers, then, had significant authority over the writing of their texts. The texts were to be accurate, reproducible, and workable, but the developers' editorial responsibilities were limited to making an effort to render their code readable by other developers. The social functions of the collaboration, after all, were carried out by versioning software that ensured no code set would be superimposed on another. Developers also were significantly rewarded for their text work. All were in a relatively low place on SecureCom's organizational chart, and most had only a bachelor's degree in computer science; some had not even that. But all were nonetheless significantly rewarded in material, social, and symbolic capital, with the amounts of capital structured by the genres in which they worked. Those writing texts closest to the guts of the machine, the farthest from human contact, received the most capital. For example, machine-code writers were better paid than GUI designers, whose work was intensely tied to the idiosyncrasies of end users instead of to the more rigid constraints of the machines themselves.

Waldo, for instance, was a machine-code developer who wrote the low-level code that determined how the product's hardware networked with its administrative machine. His work, then, was considered to be closer to the machine than that of Dean, who developed the GUI that determined how users interacted with the product's functions. As a result, Waldo was given a higher salary than Dean, who in turn still received more material, social, and symbolic capital than most others in the organization. Waldo and Dean were more limited in their freedom of movement and task selection than the COO, partly because of the necessity of their being tied to their machines all day and partly because they were limited by the requirement that they create the agreed-upon product features. Nonetheless they were allowed significant creativity in planning their schedules, taking vacations, taking smoke breaks, taking dart-gun breaks, and deciding their tasks. Waldo, for example, came and went as he pleased; he generally came in around 3 p.m., worked a few hours, and left at around 9 or 10 p.m., rarely working a full day. He frequently missed meetings and occasionally even misinterpreted or miscoded because he was not in on crucial decision-making processes. Still, when SecureCom became concerned that he might leave the company, he was offered a $10,000 raise with bonuses for showing up at work for meetings and for work during core hours of the day.

This kind of enormous social and symbolic capital was not limited to developers; even low-rung technical workers had significant power to shape their environs through creative use of their tools—in one case an e-mail signature file. Hank, the developer's intern, was not financially rewarded highly for his work, but he received other types of payoff for his perceived expertise. Hank was originally hired to assist in constructing the hardware components of the product, but he quickly tired of that and refashioned himself as the organization's system administrator, making himself responsible for network management, password administration, and machine upkeep. He made the switch by changing his e-mail signature file to say "Hank, System Administrator" and by starting to do the work; no one questioned his move. Later, when he decided he would like to be a developer, he added Developer to his e-mail signature file. Again, no one objected, and he soon added developer tasks to his daily activities. In essence, his cultural and symbolic capital as a technical person paid off in significant social capital, despite his low position on the organizational chart. Hank also was given a control over defining his genre sets (in this case, using his e-mail signature to define the reality of his position) that less technical others higher on the organizational tree could only dream of.

Empowerment and Nontechnical Workers

My point here is not just to complain about how easy the technical workers had it at my last job; it is to suggest that the technical-activity system offered particular kinds of rewards, and that those rewards varied depending on the technical genres with which one worked. But as we saw in Hank's case, even interns, if technical enough, had significant power and abilities to use their genres to alter their activity systems and to grant themselves access to different kinds of capital. At the same time, it is not the case that all nontechnical workers received fewer rewards than did the technical workers. Evan, as the organization's COO with primary responsibilities for sales and funding, was the most highly rewarded employee in the organization. He had significant power in all forms of capital. Evan wrote documents that were mission critical and that constructed the reality of the company for those inside and outside the organization. He wrote company descriptions, including business plans that detailed the current state and future goals of the company for potential investors, and corporate overviews that described and created company reality for Web site visitors and seminar attendees as well as company employees. He also wrote documents that described and created the company's relationships with the outside world in more concrete ways, including stock option statements for employees, maintenance agreements, and process documents that established and limited sales staff relationships with their customers.

Evan had substantial authority over the writing of these texts. With few exceptions that involved the input of SecureCom's board, he had total control over his shaping of the organization with these documents. The original cofounders of the organization of course had veto power, but they rarely read these documents let alone enacted that power. These documents, after all, were the reason they hired Evan; company investors, and the cofounders' sense of the corporate genre, dictated that these types of documents needed to exist. For this self-directed work, Evan was significantly rewarded, was made the highest paid employee, and was granted complete freedom to come and go as he pleased and select and organize his own activities.

That said, as with all the nontechnical workers, Evan was confronted daily with the politics of the product. He, like all of the nontechnical workers, worked with overtly political genres every day in creating new brochures or selling the product to a potential client. As a result, most of these nontechnical workers had thought about the politics of filtering a great deal. Helen spoke pointedly in her discussion of the politics.

> I have no problem with it at all. I suppose you know the key issue is always whether we should censor. I get around all those issues by saying this is a time-based technology. The world is run by economics whether we like it or not.

So the work genres assigned to the nontechnical workers did not allow them to ignore the larger politics of their product. However, unlike Evan's company-building documents, their phone calls, sales letters, and flyers were frequently less valued than more technical (and thus subtly political) texts like technical specifications and computer code. The value placed on production of the company's narrative varied along a continuum based on the percentage of their job spent bringing in money; just as technical workers were rewarded for being close to the machine, nontechnical workers were rewarded for being close to the bottom line. Those closest—salespeople—were given livable salaries in addition to receiving commissions, and they had near-complete freedom to plan their client visits and phone calls; at the same time, they lacked the security and stability granted to technical workers.

Those farthest from the money were, naturally, also the farthest from the machine, and the workers who rode the machine/money boundary most ambiguously were the organization's least rewarded. For example, Ellen the marketing intern, unlike Hank, was neither clearly technical nor directly related to the bottom line, so although she had a degree in her field of specialization and was working on a degree in MIS, she was nonetheless lower paid and had significantly less freedom of movement. She was, for example, expected to stay in her cubicle in case the secretary's phone rolled over to hers. Supply and demand is a possible explanation, but the nearby university of science and technology actually made hiring marketing specialists more difficult than hiring computer specialists; this was particularly so if marketing specialists had computer expertise. Gender is also a possible answer, although two of the principals, and nearly half of those highly rewarded with capital within the organization were women. Besides, Ellen's capital limitations were not true only of boundary-line interns; in general, boundary workers were more limited in their symbolic, material, and social capital, regardless of their cultural capital and their positions on the organizational chart.

Helen, the director of marketing, for example, brought significant cultural capital to the organization as someone with an MBA (the same degree Evan had) and previous experience as a marketing director for a software company. But while her job was certainly significantly technical, she was neither technical enough nor bottom line enough to be viewed as writing the kinds of mission-critical texts that would give her the kinds of freedom granted workers on the far ends of the spectrum. Her work was closely supervised by Evan; nothing was permitted to be sent out of the organization without his checking it for its matching up to his vision for the organization, and she often found herself fighting with him for control over the kinds of marketing tasks she wished to conduct. Her social capital, then, was limited, and her symbolic capital was constantly under question by Evan and the developers, all of whom believed that marketing was a nicety rather than a necessity. Finally, Helen's material capital also was limited; despite her high position on the organizational chart (see Figure 1) and her long hours, her pay was significantly less than that of even the lowest-paid developer.

My capital acquisition as a writer was similarly limited by my ambiguous position as a boundary worker. Half of my work—the technical documentation half—was viewed by the developers as critical to the selling of the product and as marginally technical; the four cofounders had long outsourced technical writing because they saw it as both essential and because it was a skill set they did not possess. At the same time, my work was frequently nontechnical; my job was to design or write many of the texts not written by those on the other ends of the spectrum, from white papers to marketing brochures. The strange combinations of genres I wrote on a daily basis forced me, like many technical writers, to straddle the boundary between the technical and nontechnical worlds, receiving the complete capital benefits of neither. While my work was rewarded with greater material capital than Helen's, and I gradually negotiated freedom to come and go, I was still paid less than the developers, had little freedom to choose my

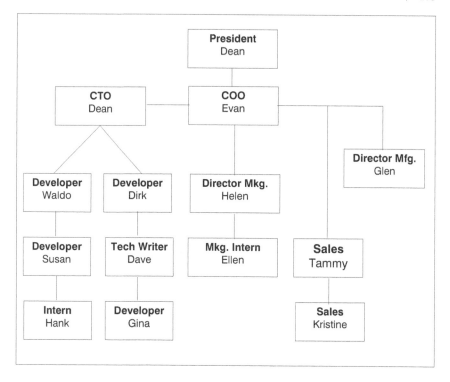

Figure 1. Abbreviated organization chart.

tasks, and, as with other boundary workers like the marketers, my less-than-mission-critical activities were subject to intense observation and scrutiny from both ends of the spectrum.

For example, one of Evan's first moves as manager was to implement a content-management system that allowed for source control of word-processed documents instead of computer code. The system allowed writers to access their files from anywhere, track multiple revisions, and quickly and easily access each other's work. It also allowed for simple and easy supervision of writing work; Evan would daily check the "what is new" page within the site to see what had been accomplished, and he subscribed to a number of my documents so that he would receive e-mail notifications when changes were made, allowing him a panoptic view of boundary genres from anywhere with a Web browser. Nontechnical work was equally subject to the suggestions of those on the technical side who went over each document rigorously to ensure it was technically correct and thus in keeping with their own mission-critical texts. Nontechnical workers could only dream of being granted the kinds of genre authority granted to Hank.

CONCLUSION

After I left the company, SecureCom eliminated another technical/nontechnical slippage by hiring a designated technical writer on contract, allowing her high cash reward but virtually no security or stability, and handing all the overlapping marketing materials to Ellen, who was hired on full time and gradually permeated the boundary enough to take control of the Web site. Activity systems are, in Catherine Schryer's (1993) famous phrase, always only "stabilized for now" (p. 204), allowing for the possibility that an actor's use of their tools might lead to change in the activity system. For example, Ellen, through training, was able to gain enough technical expertise to be given the task she wanted. Was she empowered, then, in ways that would not have happened to a Taylorist assembly-line worker or typist? It is hard to say. Working on the Web site afforded her some additional symbolic capital because everyday something she did could be seen by the public. However, it did not get her a raise or any additional decision-making authority, and in fact it put her under more scrutiny from both technical and nontechnical staff; it was a mixed bag.

So what is my point? I can now argue that Ellen and other knowledge workers at SecureCom cannot be said to have been empowered, using my definition that incorporates the individual perspective, financial rewards, value, and authority placed on work. But I argued that point in my opening; the utopian empowerment narratives are already fairly transparent and obvious to those of us without a stake in maintaining them. And the point is not simply that workers are more and less empowered than each other, which of course we already suspected, but to investigate the means by which that empowerment is defined, created, accessed, and regulated. By following the cultural turn, by incorporating postsocial constructionist theories such as activity theory and genre theory, and by relying on valuable cultural work like Bourdieu's, we can create richer descriptions and assessments of how workers can or cannot access decision-making authority in their work.

The important work here is to create more nuanced discussions of power than that provided by empowered/disempowered and similar binary approaches to culture. Using a cultural approach, we can build understandings of organizational power that do not view empowerment as a simple hierarchy but that seek instead to understand the broad array of discourses, technologies, professions, traditions, and capital that regulate the ways that workers receive financial rewards and build or lose the authority and value attached to their work. Such an approach can easily go beyond the limited discourse of workplace empowerment, which although it is pervasive and resilient, is for our purposes easily shot down by qualitative data or even anecdotal evidence and ultimately serves here only as an extended example. But, as I hope I have indicated here, the same structures that enforce organizational access to empowerment also structure the division of tasks, organizational structure, and day-to-day work life; and further investigation of these structures as regulatory features can help us better articulate the interactions, power, and networks of discourse that make up organizational life.

REFERENCES

Agar, M. (1980). *The professional stranger: An information intro to ethnography.* Orlando, FL: Academic Press.

Aronowitz, S., & Difazio, W. (1994). *The jobless future: Sci-tech and the dogma of work.* Minneapolis: University of Minnesota Press.

Bazerman, C. (2003). What is not institutionally visible does not count: The problem of making activity assessable, accountable, and plannable. In D. Russell & C. Bazerman (Eds.), *Writing selves/writing societies: Research from activity perspectives.* Fort Collins, CO: WAC Clearinghouse, 2003. Retrieved from http://wac.colostate.edu/books/selves_societies/>.

Blyler, N., & Thralls, C. (1993). *Professional communication: The social perspective.* Newbury Park, CA: Sage Publications.

Bonnell, V., & Hunt, L. (1999). *Beyond the cultural turn: New directions in the study of society and culture.* Berkeley, CA: University of California Press.

Bourdieu, P. (1987). *Distinction: A social critique of the judgement of taste.* (R. Nice, Trans.) Boston, MA: Harvard University Press.

Bowker, G., & Star, S. L. (2000). *Sorting things out: Classification and its consequences.* Cambridge, MA: MIT Press.

Boyle, J. (1997). *Shamans, software, and spleens: Law and the construction of the information society.* Boston, MA: Harvard University Press.

Bruffee, K. (1984). Collaborative learning and the conversation of mankind. *College English, 46,* 635-652.

Clark, D. (2001). *A rhetoric of boundaries: Living and working along a technical/ non-technical split.* Unpublished doctoral dissertation, Iowa State University.

Cole, M., & Engeström, Y. (1993). A cultural-historical approach to distributed cognition. In G. Solomon (Ed.), *Distributed cognitions: Psychological and educational considerations* (pp. 1-46). Cambridge, UK: Cambridge University Press.

Davenport, T., & Prusak, L. (1998). *Working knowledge: How organizations manage what they know.* Boston, MA: Harvard Business School Press.

Dertouzos, M. (1997). *What will be: How the new world of information will change our lives.* New York: HarperEdge.

Doheny-Farina, S. (1996). *The wired neighborhood.* New Haven, CT: Yale University Press.

Downey, G. (1998). *The machine in me: An anthropologist sits among computer engineers.* New York: Routledge.

Drucker, P., Garvin, D., Dorothy, L., Straus, S., & Seely Brown, J. (1998). *Harvard business review on knowledge management.* Boston, MA: Harvard Business School Press.

Dyson, E. (1998). *Release 2.1: A design for living in the digital age.* New York: Broadway Books.

Engeström, Y. (1993). Developmental studies of work as a testbench of activity theory: The case of primary care medical practice. In S. Chaiklin & J. Lave (Eds.), *Understanding practice: Perspectives on activity and context* (pp. 64-103). Cambridge, UK: Cambridge University Press.

Flower, L. (2003). Intercultural knowledge building: The literate action of a community think tank. In D. Russell & C. Bazerman (Eds.), *Writing selves/writing societies: Research from activity perspectives.* Fort Collins, CO: WAC Clearinghouse, 2003. Retrieved from http://wac.colostate.edu/books/selves_societies.

Gates, B. (1995). *The road ahead.* New York: Viking.

Gee, J., Hull, G., & Lankshear, C. (1996). *The new work order: Behind the language of the new capitalism.* New York: Westview Press.

Greenbaum, J. (1998). From Chaplin to Dilbert: The origins of computer concepts. In S. Aronowitz (Ed.), *Post-work: The wages of cybernation.* New York: Routledge.

Hayes, R. (1995). Digital palsy: RSI and restructuring capital. In J. Brook & I. A. Boal (Eds.), *Resisting the virtual life: The culture of politics and information.* San Francisco, CA: City Lights Books.

Herndl, C. (1993). Teaching discourse and reproducing culture: A critique of research and pedagogy in professional and non-academic writing. *College Composition and Communication, 4,* 349-361.

Hull, G. (1999). What's in a label? Complicating notions of the skills-poor worker. *Written Communication, 16,* 379-411.

Hutchins, E. (1993). Learning to navigate. In S. Chaiklin & J. Lave (Eds.), *Understanding practice: Perspectives on activity and context* (pp. 35-63). Cambridge, UK: Cambridge University Press.

Hutchins, E. (1996). *Cognition in the wild.* Cambridge, MA: Massachusetts Institute of Technology Press.

Latour, B. (1991). Technology is society made durable. In J. Law (Eds.), *A sociology of monsters: Essays on power, technology, and domination* (pp. 103-131). London: Routledge.

Longo, B. (1998). An approach for applying cultural study theory to technical writing research. *Technical Communication Quarterly, 7,* 53-73.

Myers, G. (1996). Out of the laboratory and down to the bay. *Written Communication, 13,* 5-43.

Marvin, C. (1990). *When old technologies were new: Thinking about electric communication in the late nineteenth century.* Oxford: Oxford University Press.

Negroponte, N. (1995). *Being digital.* London: Hodder & Stoughton.

Pavlicek, R. (2000). *Embracing insanity: Open source software development.* New York: Sams Publishing.

Peters, T. (1994). *Liberation management: Necessary disorganization for the nanosecond nineties.* New York: Ballantine.

Raymond, E. (2001). *The cathedral and the bazaar: Musings on Linux and open source by an accidental revolutionary.* Sebastopol, CA: O'Reilly.

Rifkin, J. (1995). *The end of work: Technology, jobs, and your future.* New York: Tarcher/Putnam.

Russell, D. (1997a). Writing and genre in higher education and workplaces: A review of studies that use cultural-historical activity theory. *Mind, Culture and Activity, 4,* 224-237.

Russell, D. (1997b). Rethinking genre in school and society: An activity theory analysis. *Written Communication, 14,* 504-554.

Schryer, C. F. (1993). Records as genre. *Written Communication, 10,* 200-234.

Senge, P. (1990). *The fifth discipline: The art and practice of the learning organization.* New York: Doubleday Press.

Standage, T. (1999). *The Victorian Internet: The remarkable story of the telegraph and the nineteenth century's online pioneers.* Berkeley, CA: Berkeley Publishing Group.

Tapscott, D. (1995). *The digital economy: Promise and peril in the age of networked intelligence.* New York: McGraw-Hill.

Tenner, E. (1997). *Why things bite back: Technology and the revenge of unintended consequences.* New York: Vintage.

Ullman, E. (1997). *Close to the machine: Technophilia and its discontents.* San Francisco, CA: City Lights.

Winsor, D. (2003). *Writing power: Communication in an engineering center.* Albany: State University of New York Press.

CHAPTER 9

Power as Interactional Accomplishment: An Ethnomethodological Perspective on the Regulation of Communicative Practice in Organizations

Barbara Schneider

I propose an ethnomethodological approach to understanding power that offers insights into the regulation of communicative practice in organizations. I briefly review functionalist and critical perspectives on power and regulation and then elaborate a view of power as an interactional accomplishment produced by participants in the course of social interaction. This approach offers a way to see that neither the rules of particular genres of discourse nor the power of particular organization members determine individual communicative activities. Rather participants orient to those rules and power relations and actively reproduce versions of them in their interaction. I advocate an approach to studying power and the regulation of communicative practice through the study of naturally occurring social interaction and illustrate with an analysis of two short examples.

The issue of the relationship between the larger social context and communicative practices in organizations has become a central concern among scholars in a number of disciplines. The debate focuses on how aspects of context regulate the communicative activities of individuals in organizations and how these individual activities in turn influence organizational context. In this chapter, I contribute to this debate with an approach to understanding power inspired by the work of Harold Garfinkel (1967a) and his followers in the field of ethnomethodology. I suggest that power be regarded as a practical achievement, produced by participants in the course of social interaction in which some versions or accounts of "reality" come to dominate others. I believe that such an understanding of power can lead to a deeper understanding of the relationship between social context and individuals' communication activities and offer insights into the regulation of communicative practice in organizations.

Reed (1996) describes power as "the most overused and least understood" (p. 40) concept in organization studies. This is hardly surprising given the pervasive but contingent, elusive, and ephemeral nature of the phenomenon we experience and refer to as power. People, groups, and institutions can seem to have it one moment and lose it the next, but somehow we can never manage to say what it really is. The very language we use to talk about it leads us to regard it as a thing of some kind—power can be held, used, manipulated, wielded, exercised, transferred, shared, delegated, increased, lost—but this does not help us to define its essence. At the risk of taking on a concept that others have been unable to pin down or agree upon, I would like in this chapter to propose a view of power as an interactional accomplishment, produced by participants in the course of social interaction. Power, in this view, is not a discrete entity that is somehow conferred on people or offices by virtue of their position in an organizational hierarchy. It is, rather, something that must be accomplished over and over again in every social interaction by those who would say and have it said about them that they have power. This perspective moves communication and communicative practices into central focus in the understanding and study of power. Rather than trying to describe power as something that is presumed to exist as a discrete and recognizable entity in the world and to describe its workings, I suggest that a more productive way to understand power, and through this, the regulation of communicative practice, is to examine its accomplishment in everyday social interaction through the study of communicative activity.

I begin by briefly describing a number of perspectives on power, both traditional and more recent. I then outline an ethnomethodological perspective on power, with particular attention to the way in which this approach addresses the structure/agency debate, a central question in understanding the regulation of communicative practice. I then discuss ethnomethodological studies of communication in organizational settings and close with an analysis of two brief examples of social interaction from my own research.

EARLY PERSPECTIVES ON POWER

Although it is beyond the scope of this chapter to survey in detail the very large literature on power in organizations, I draw on excellent reviews found in Hardy and Clegg (1996), Reed (1996), and Mumby (2000) to provide a brief and somewhat oversimplified outline of several threads of theory and research in the area. Hardy and Clegg identify two primary "founding voices" (p. 623) in the study of power: the functionalist and the critical, each of which has led to large, but not overlapping, bodies of literature on power. They also identify a more recent third, Foucauldian, voice in the study of power. These perspectives make very different assumptions about the nature of power and imply very different views of the regulation of communicative practice.

In the functionalist perspective, the founding voice is a managerial one, in which organizational interests are equated with managerial interests. The hierarchical distribution of power within organizations is regarded as legitimate, and the use of power by those who are not "supposed" to have it (e.g., lower-ranking organization members) is viewed as illegitimate and dysfunctional. This perspective takes for granted the authority of organizational elites and assumes that managers use power in responsible ways to achieve organizational goals, while others use it irresponsibly to challenge and resist these goals. Implicit in this perspective is a view of regulation, whether of communicative practice or other aspects of employees' behavior, as a legitimate function of managers, an essential tool for advancing organizational goals. Important contributors to this stream include French and Raven (1959), whose analysis of the bases of power is still widely cited today; the Carnegie group (e.g., Cyert & March, 1963; Simon, 1976); and theorists in the areas of strategic contingency (e.g., Hickson, Hinings, Lee, Schneck, & Pennings, 1971) and resource dependency (e.g., Salancik & Pfeffer, 1977).

The founding voice for the critical perspective is rooted in the work of Marx and Weber. It focuses on the existence of conflicting interests in organizations and studies power as domination. In this view, power is derived from ownership and control of the means of production. Organizational structures and rules distribute and legitimate power in ways that serve the interests of some but not other groups within the organization. Although those lower in the hierarchy have some room to resist the power that controls them, dominant groups generally have access to resources that make it difficult for lower ranking groups to mount successful challenges. Implicit in this perspective is a view of regulation as a structural phenomenon, embedded in hierarchical power relations. Important contributors to this stream of research include Bachrach and Baratz (1962), who studied power as it was evident in decision making in organizations; and Lukes (1974), who extended the work of Bachrach and Baratz, pointing out that power must also be seen in the way that the interests of some groups remain unarticulated. Mumby (2000) points out that these early voices focus on "the cognitive, decision-making and structural" (p. 613) aspects of power, with communication and discourse ignored or relegated to a simple transmission role.

A third and more recent view of power is that provided by Foucault (e.g., 1977, 1980) and those who have applied his ideas to organizational settings. This perspective calls into question the sovereign notion of power embedded in both the functionalist and the critical perspectives (i.e., that power is imposed from above) and asserts that power "is embedded in the fiber and fabric of everyday life" (Hardy & Clegg, 1996, p. 631). Power is no longer a resource, held in varying degrees by organization members and used to affect the behavior of others. It has no essence and cannot be described or measured. Power resides instead in what Foucault (1980) calls discursive formations, historically and culturally located systems of power/knowledge. Discursive formations provide interpretive frameworks that organize understanding of particular social settings and relationships. They are characterized by rules that regulate who is allowed to speak, what can be spoken about and how, and what kinds of things will count as legitimate knowledge. In this view, power is widely dispersed, embedded in the network of relationships in organizations. It is not simply a negative force, used to control and repress others; it is a positive force, producing knowledge, identities, and possibilities for behaviors and actions in particular settings.

Foucault's ideas shift the focus in the study of power and regulation in organizations and open the possibility of understanding power in organizations through a communicational lens. Foucault's ideas have been taken up by a number of recent voices including organizational communication scholars in the critical tradition (e.g., Deetz & Mumby, 1990; Mumby, 2000; Mumby & Stohl, 1991) who propose an explicitly communicational view of organizations and of power in organizations. Mumby (2000), for example, describes communication, organization, and power as interdependent phenomena in which the production of organizational meanings through communicative practice is seen as "fundamentally mediated by power" (p. 585). Foucault's ideas have also been incorporated into the ethnomethodological tradition in the work of a number of scholars (e.g., Gubrium & Holstein, 2000; Hutchby, 1996; Miller, 1997b) who have noted striking parallels between Foucault's ideas and the ethnomethodological interest in how language and social interaction constitute social life. Foucault himself, however, paid very little attention to specific instances of social interaction, focusing instead on historical examinations of the larger workings of discursive formations. Ethnomethodology, on the other hand, offers not only a conceptual framework for understanding power and the regulation of discourse in organizational settings but also a specific approach to the study of everyday communication practices. I believe that ethnomethodology has much to contribute to an explicitly communicational view of organizations and offers a conceptualization of power that is particularly useful in understanding the regulation of organizational communicative practice. I therefore turn now to discussion of an ethnomethodological perspective on power.

AN ETHNOMETHODOLOGICAL UNDERSTANDING OF POWER

As is evident in the brief review above, organizational scholars have defined power in a number of different ways. Rather than wading in to make another attempt to define power, ethnomethodology suggests instead that power should be understood from the point of view of participants in social settings. The questions to be answered then are not what is power and how does it work, but rather how do participants accomplish what count as power relations for them in particular social settings. This question is embedded in a larger ethnomethodological concern for the social and communicative practices through which "participants create, assemble, produce and reproduce the social structure to which they orient" (Heritage, 1987, p. 231). Ethnomethodologists are interested in the methods through which setting members accomplish a sense of social reality and "confer privilege" (Holstein & Gubrium, 1994, p. 264) on some versions of reality rather than on others. In ethnomethodological lingo, this is often described as studying how participants "do" social life.

Ethnomethodology grew out of the foundational work of Harold Garfinkel (1967a), who insisted that all social order is organized from within the social situation by members participating in "a stream of experience" (Boden, 1994, p. 46) at a specific time and in a particular location. He was responding to the dominant sociological view at the time, promoted by the work of Talcott Parsons, that social order is possible because individuals internalize and then act out existing rules, norms, and values. Rather than seeing rules and norms as causal explanations for individual behavior, Garfinkel was interested instead in how individuals call on rules and norms to make their behavior accountable and in doing so reproduce those rules as social "facts." He regarded the apparently stable external world as constituted through the situated reasoning of knowledgeable actors as they accomplish their practical purposes. He called for researchers to understand the taken-for-granted social world as an ongoing accomplishment of actors' interpretive work and for research to focus on, as Brandt (1992) says, how people "make circumstances look as if they aren't created at all, but are simply there for everybody to see" (p. 319). This interest in the *hows* of social life, that is, in the methods through which setting members actively produce and sustain social settings, has been the driving interest of ethnomethodological research. It has led to a particular focus on the study of naturally occurring social interaction, that is, on the communicative activities that take place between setting members (rather than with or for researchers) in the course of carrying out their daily practical purposes. In other words, ethnomethodological researchers regard the social interaction of setting members itself as the topic of study rather than as a resource for information about various aspects of their lives.

While there are many aspects of ethnomethodology that could be discussed here, I focus on one that is of particular relevance to understanding power and the regulation of communicative practice in organizations. This aspect is the ethnomethodological contribution to understanding the nature of social structure and the link between social structure and human action. These questions have been hotly debated in various bodies of literatures in what is known as the agency/structure or micro/macro debate. This debate seeks to clarify the relationship between larger social structures and the behavior of individuals living within those social structures. The term "social structure" has been used in many ways (Schegloff, 1991), but it is generally used to refer to stable patterns of social relationships that are external to individuals (Schegloff, 1991; Wilson, 1991; Zimmerman & Boden, 1991). It is a large-scale, or macro, arrangement that provides the context for all human activities, or microphenomena. The link between the domains of social structure and human activity is variously regarded as causal, with local human activities being seen as a product of larger social forces, as mutually constraining or enriching, or as the interaction of various levels of social life (Hilbert, 1990).

For ethnomethodologists, the agency/structure distinction is a false dichotomy. Neither affirming nor denying the existence of social structure, ethnomethodologists are interested instead in the question of how social order is produced as a local accomplishment of situated actors. Structure is realized through the actions of agents, and therefore the two cannot be studied as independent domains that interact in particular ways. Rather than studying the "objective reality of social facts" (Zimmerman & Pollner, 1971, p. 81), ethnomethodologists seek to investigate how social facts are assembled as an apparently objective reality. They investigate the local social practices and interactions through which structure is "made to happen, made to appear" (Hilbert, 1990, p. 795). If social structure is to be studied at all, it is studied as "something that humans do, rather than something that happens to them" (Boden, 1994, p. 11). Wilson (1991) proposes a definition of social structure that focuses on the immediate experiences of actors in social interaction:

> Social structure consists of matters that are described and oriented to by members of society on relevant occasions as essential resources for conducting their affairs and, at the same time, reproduced as external and constraining social facts through that same social interaction. (p. 27)

That is, social structure is produced and reproduced as circumstances that are apparently external to individuals' activities in specific social settings through the very activities in which individuals talk and behave as though that structure were external.

Social actors are thus seen not as pawns, moved around at will by forces in the social environment within which they happen to find themselves. Rather, they are regarded as reflexive beings, "active agents in the constitution of their unfolding social worlds" (Boden, 1990, p. 203). They are knowledgeable actors who attribute meanings to their joint actions that both shape and renew their understandings of their social worlds. This does not mean that individuals do not experience social structure as a social fact prior to and separate from their own participation in a particular social setting. However, as they orient to the social world as external and constraining, they reproduce this social world as an apparently objective reality. Their activities are both embedded in and constitutive of their social environment.

If we apply these ideas to organizations, we come to a very different understanding of organizational power and the regulation of organizational communication than that offered by traditional approaches. Traditional approaches typically regard organizations as structured entities that have an existence separate from the individuals who work in them. Ethnomethodologists, on the other hand, regard organizations not as externally constraining structures but as the "observable and recurring activities through which members construct social settings, relationships, and realities" (Miller, 1994, p. 287). This shifts the focus of study from the apparently real abstract structures of organizations to the communicative activities through which those structures are constructed and maintained. Miller (1994) regards organizations as "situated conventions" (p. 282) that provide conditions of possibility for interpretive activity. These conventions make available certain preferred ways of thinking and acting within the organization and repress other less preferred ways of thinking and acting (Miller, 1994) and make it likely that some but not other readings, or interpretations, of the organization will be regarded as legitimate by others in the organization. However, interpretive conventions do not prescribe behavior for people in organizations. People construct warrantable arguments for their versions of the organization, using the organization's preferred modes of interaction to "make some reality claims more available and credible than others" (Miller, 1994, p. 290). Interpretive conventions thus shape but do not determine participants' thoughts and actions. They provide a context within which individuals assimilate and organize experience into apparently stable social facts.

People in organizations use the interactional and interpretive conventions available to them to construct, among other aspects of the setting, the power relations of the organization. In this view, power and dominance are not structural features of organizational life and the regulation of communicative practice is not the product of structural forces. Instead, these are accomplished by organization members in social interaction as they pursue their practical interests in the organizational setting. The apparently objective realities of power relations and regulation are then, to call on Wilson's (1991) definition, something to which organization members orient in carrying out their affairs within the organization and, in doing so, reproduce. That is to say, organization members experience the social structure and power relations of the organization as social facts that exist prior to and separate from their own membership in the organization, and in orienting to these as social facts, reproduce them as external realities.

In this view, social settings are never settled once and for all; they are constantly shifting, constantly accomplished in social interaction. Even when the conventions of an organization seem settled, as L. Miller (1993) says, "every apparently settled order conceals a reality struggle" in which one account of the organization has come to dominate all other possible accounts. In this view, power can be understood as a practical achievement in social settings, in which some voices dominate other voices, some points of view prevail in being considered better or more appropriate for the organization. In short, power might be seen as a question of whose account (of the organization or in a particular situation) counts.

Ethnomethodological Studies of Organizations

In the years since Garfinkel's (1967a) early work, an extensive literature studying organizations from an ethnomethodological perspective has developed. This literature includes studies of legal settings (e.g., Atkinson, 1992; Holstein, 1988), educational settings (e.g., Mehan, 1978), medical settings (e.g., Anspach, 1987; ten Have, 1991), long-term care facilities (e g., Gubrium & Buckholdt, 1979), social service agencies (e.g., Holstein, 1992; Zimmerman, 1969), emergency call centers (e.g., Whalen, Zimmerman, & Whalen, 1988), news interviews (e.g., Clayman, 1988), job interviews (e.g., Button, 1992; Silverman & Jones, 1973), women's shelters (Loseke, 1989, 2000), and so on. It would be misleading to present this research as a cohesive body of work (Atkinson, 1988), as these studies represent a variety of analytic approaches inspired by the tenets of ethnomethodology. Nevertheless, all focus on the study of how the social realities of organizational settings are constructed through language use and social interaction among setting participants.

The word "power" is rarely used in the ethnomethodological literature; however, there is a tacit understanding that power is accomplished by those whose versions of social reality become accepted. In the settings described in these studies, it would be easy to fall back on a reified notion of power and say that social interaction unfolds as it does because some participants *have* more power than others and therefore have the ability to control the course and outcome of interaction. However, an important tenet of ethnomethodological research is that concepts such as power cannot be invoked as explanations for social phenomena unless evidence can be found in the details of social interaction that participants themselves orient to those particular aspects of social context as relevant in that social interaction. As Watson (1990) says, power (or any other explanatory concept) "must be firmly located in the systematic examination of features integral to the discourse itself" (p. 280). In this view, conclusions about the regulation of communicative practice must come from the study of the details of social interaction rather than from generalizations about social context derived from sources outside the interaction.

One of the few to focus directly on the issue of power is Hutchby (1996), a conversation analyst who says that the "ways in which participants design their interaction can have the effect of placing them in a relationship where discourse strategies of greater or lesser power are differentially available to each of them" (p. 482). As discourse unfolds, some participants in interaction have access to, or make available to themselves, interactional resources not available to others and therefore can achieve effects not achieved by others. He examines a radio talk show in which people call in to the show to discuss topics of their choice. This particular show provides a context in which callers typically introduce a topic and express an opinion on it; that is, they take the first turn in the conversation. This lets the host go second, a position from which the host can challenge callers without revealing his or her own opinions, thus giving the host the upper hand in the conversation. Because of differential access to the second position, the host is easily able to put callers on the defensive and thus appears to "have" more power than the callers. From time to time, callers turn the tables by asking questions of the host, thereby putting themselves into the more powerful second position, even if only briefly. Hutchby shows that we do not have to assume that the host has more power simply because of his or her position as host and reputation as a celebrity. Instead, power is accomplished in interaction through differential access to interactional resources and can be observed in the details of specific interactions that reproduce the host's better access to these resources. The course of the interaction is regulated not so much by preexisting power relations as by the way participants themselves orient to the context and design their interaction.

ANALYZING NATURALLY OCCURRING DATA

To take a closer look at how power and regulation are accomplished in social interaction, I now turn to an analysis of two short episodes of social interaction that took place during the construction of an organizational text. As noted above, ethnomethodology has produced a number of analytic approaches. I use an approach here that I believe is particularly valuable in studying the regulation of communicative practice in organizations. This is the approach of constitutive ethnography, proposed by Mehan and used by scholars such as Holstein (1993) and G. Miller (1991). This method puts "structure and structuring activities on an equal footing by showing how the social facts of the world emerge from structuring work to become external and constraining, as part of a world that is both of our making and beyond our making" (Mehan, 1978, p. 60). This method has been developed further with the addition of a Foucauldian perspective by G. Miller (1994, 1997b) and Gubrium and Holstein (2000). Gubrium and Holstein's (2000) analytics of interpretive practice focuses attention on "how members artfully put discourses to work as they constitute their . . . social worlds" (p. 497). This approach emphasizes close attention to the specifics of social interaction, what they call the "hows" of reality construction, while not ignoring the larger "whats" of social context. This focus on the interplay of individual communicative activity and larger social and organizational discourses allows researchers to study the communicative practices of individuals in organizations as simultaneously embedded in and constitutive of the social reality of organizations without privileging either structure or agency.

The conversations analyzed here were recorded as part of a much larger study in which I observed two managers in a music conservatory situated within a community college conduct an internal administrative review and evaluation of a group of their educational programs for young children. (For a full description of this project, please see Schneider, 2000, 2001, 2002.) In carrying out their review and evaluation, the two managers interviewed all the teachers in the programs, took notes on the interviews, compiled the notes, and then wrote a report. The data for the study consisted of transcriptions of tapes of meetings between the managers, transcriptions of the interviews the managers conducted with teachers, notes the managers took during the interviews, compilations of the notes, and four drafts of the report.

The examples discussed here come from interviews with two teachers (the first, female, and the second, male) conducted by one of the managers. My analysis of the first example shows that the manager and the teacher produce discourse in which the manager not only maintains control over the flow of the conversation but actively participates in the production of the teacher's answers. The manager and teacher draw on conditions of possibility provided by the context and collaborate to produce interaction in which certain discourse strategies are more available to the manager than to the teacher, and through this, reproduce the existing power relations. This conversation may seem to be rather insignificant, but it illustrates the potential for power and regulation to be produced in any interaction, no matter how slight.

Example 1

Manager: Why do you think people come to Community College?

Teacher: Because it has a very good reputation.

Manager: The College? The Conservatory? The Early Childhood [program]?

Teacher: I would say the Conservatory, the music program, has a really good reputation.

Manager: Let's say you are hypothetically a parent of young children. And there are a lot of different things you can do. You can go to swim and gym, you can go to ballet, lots of things.

Teacher: It is also probably that it is a college and it is established and offers a wide range of educational possibilities.

In this example, the manager first asks his question as it is printed on his interview schedule. The very asking of the question places the teacher in the position of having to produce a response that will be understood as an answer to that question. If she does not conform to this expectation, she will likely be regarded as socially or professionally incompetent. The answer the teacher gives contains the pronoun "it." Although it is probably clear to both manager and teacher that she is referring to the conservatory, the manager nevertheless asks in his next question for specification of a referent. He offers her three choices from which to construct her response, thus constraining her answer. She selects one of the choices, saying that the reputation of the conservatory attracts students. She thus cooperates with him to limit her answer to the choices he offered, accepting the constraints imposed by his phrasing of the question.

However, even this does not seem to be a complete enough answer because the manager now reframes the question and repositions her, asking her to imagine herself the parent of young children (which she, in fact, is) rather than a teacher in the program: "Let's say you are hypothetically the parent of young children." By reframing the question, the manager has communicated his assessment that the teacher's first two answers have not adequately addressed the question, and the teacher responds by producing a different answer than she did before the manager assigned a new identity to her. That is, he actively encourages the teacher to produce an entirely different answer, and the teacher cooperates by doing so. She now says that it is community college that attracts students. This response is not her only possible option in answering the question. She could, for example, protest that she has already answered the question, say that her first answer is the one she is sticking with, or turn the tables and ask the manager why *he* thinks people come to community college. On the contrary, she allows herself to be repositioned by the manager and obligingly produces another answer, different from her first. She cooperates in her own repositioning by the manager and collaborates with him to produce interaction in which he has the right to reposition her.

In this exchange, it almost seems that the manager is fishing for an answer; that is, he has a particular answer in mind and does not stop asking until he gets it. In fact, at a later meeting when the two managers discuss the interviews, he reveals that this may indeed have been so. The manager who conducted the interview asks his colleague the following question: "Does that surprise you at all—that they feel that community college has a cachet that draws people but not the conservatory?" It seems that he has noticed a pattern in the answers to this question. Perhaps he has even, unknowingly, produced the pattern. Certainly, in example 1, he has participated in the production of the answer he is looking for. Before the manager repositions the teacher, she in fact says that the conservatory's reputation draws people. That is, she gives the answer that he claims (in his question to the other manager) most of the teachers do not give. Her answer, the one she gives after being repositioned, is now amalgamated in his mind with other teachers' answers to this question and will be used as the basis for conclusions and recommendations in the final report.

The conversation between the manager and the teacher took place within the context of an interview. The right to ask questions generally provides the questioner, in this case the manager, with a powerful source of control over interaction. As Sacks (1992) points out, "As long as one is in the position of doing the questions, then, in part, one has control of the conversation" (p. 54). Individual questions place immediate constraints on the discourse options available to the teacher at any given moment in the interview, and the sequence of questions allows the manager to keep the focus of the talk throughout the interview on his practical concerns. The teachers do from time to time during the interviews ask questions themselves, demonstrating that the interview setting does not give only managers the right to ask the questions. Power does not simply inhere in the managers by virtue of their position in the organization. Rather, the manager and teacher actively collaborate to produce interview discourse in which the manager asks the questions and the teacher answers them. The teacher cooperates in producing the manager's control over the course of the conversation—what Briggs (1986) calls the "communicative hegemony" (p. 123) of the manager. The teacher collaborates in the production and maintenance of a power relation in which the manager has control over the questions that will be asked and how they will be asked and therefore over the topics that will be discussed and how those topics will be discussed.

Analysis of this example shows that it is not necessary to invoke the rather vague, messy, and ill-defined concept of power in order to explain what might be regarded as the manager's regulation of this interaction. The manager accomplishes his practical aims in the interview and thus seems in the traditional view to have more power than the teacher, not because of their respective places in the social structure of the institution but because the manager and the teacher collaborate to produce interaction in which the manager has access to interactional resources that enable him to achieve certain outcomes. Even if it seems that the teacher collaborates with the manager in ways that may seem contrary to her own best interests, we do not need to fall back on a vague idea of power to explain this. The hierarchical power structure of the organization and the genre of the interview, with its rules about who asks and who answers questions, do not regulate the interaction so much as they provide a context to which both teacher and manager orient, thereby reproducing the context in their interaction.

In the example I have just analyzed, power is produced in a rather predictable way—the manager controls the flow of the conversation and the teacher collaborates to ensure that this is the case. In the next example, the teacher does not collaborate with the manager and does not allow himself to be positioned by the manager's questions. Neither the manager's position in the organizational hierarchy nor the conventions of the interview situation provide the manager with the power to make the teacher give the answers that the manager seems to be looking for. This example also illustrates the way in which any particular interaction is embedded in the ongoing stream of communicative interaction in an organization and is both shaped by larger discourses and conditions of possibility and contributes to or resists those discourses. This second example was reconstructed from notes taken during a five-minute period at the beginning of the interview when the tape recorder failed to function.

Example 2
Manager: What in your view are the particular strengths and weaknesses of the early childhood program?

Teacher: We have an excellent program. . . . We have two excellent instructors. I don't know of any weaknesses. If I did, I would be fixing them.

Manager: Anything we can do better?

Teacher: Nope. As I say, if there were any problems, I would be fixing them.

Manager: Are there any opportunities we are not taking advantage of?

Teacher: No, none.

In this brief exchange, we can see the teacher responding very differently to the manager's questions than did the teacher in the first example. In his initial question, the manager asks the teacher to describe both strengths and weaknesses. In his response, the teacher talks of strengths and claims to know of no weaknesses. As in example 1, the manager rephrases his follow-up question and thereby communicates his assessment that the teacher's answer has not adequately addressed the question. This time, however, the teacher does not cooperate with the manager. The teacher hears this question as a request for him to talk about problems and continues in this and in his next answer to reject any suggestion that the program might be in need of improvement. It seems that he is resisting the efforts of the manager to get him to talk about weaknesses of the program. In the interview, the manager then went on to talk about other topics, in effect giving up, at least for the moment, his attempt to get the teacher to identify problems in the program.

But the story, of course, does not end there. The interview is not an isolated event in the life of the organization, but is embedded in a larger network of ongoing social interactions. In a discussion between the two managers as they review a draft of the report, it becomes apparent that the managers have a very different story to tell about the program than the teacher. According to the managers, the program has serious problems, reflected primarily in rapidly declining enrollment. In fact, discussions had been ongoing between the manager who conducted the interview and the teacher about these problems and the potential for drastic changes to or even elimination of the program. We can now see both the manager and the teacher orienting to but not articulating these possibilities during the interview—the manager perhaps wanting evidence to justify cutting the program and the teacher certainly resisting giving any such evidence.

In their later discussion, the two managers agree that the teacher knows that the report they are writing will go to the academic VP. They believe that the teacher has therefore designed his answers to the interview questions to advance his practical concern of saving his program. They cast him as someone with an agenda to promote and therefore as someone whose comments about the program cannot be seen as objective. So they construct a version of the teacher that will let them discount his positive comments in the interview and instead promote their own version of a program in trouble. And indeed, before the report was finished and released, the decision to reduce the program had been made but not yet announced. The following year, the program was cancelled altogether.

It would be easy to see this outcome as the action of managers who have power. The teacher resists the manager's attempts to produce power in the moment of the interview, but it is all for naught because the managers still have the power to make the decision to cancel the program. The managers, however, are also embedded in the ongoing stream of social interaction in the college and cannot simply make any decisions they see fit. Their decisions must be approved by the academic vice president of the college. They know their report will go to others in the college and must therefore present a persuasive rationale for any program changes they propose. Just as the teacher tries to make his account of the program count in the interview, the managers must make their account of the program count within the college. Even if the managers were in a position in the college to make decisions without approval from superiors, they would not simply have the power to do so. If, for example, their decision to cancel the program was met by a huge outcry from other teachers, students in the program, and the general public with demands for meetings with the managers, for proposals to keep the program financially viable, and so on, it is possible that the outcome might have been different. If the protesters can promote their account of the program as valuable in spite of problems and thereby manage to save it, they can be seen to have accomplished power in the situation. Indeed, the deafening silence that meets many organizational decisions must also be seen as an interactional accomplishment. Organization members orient to managers as having the power to make decisions they think are appropriate for their programs and, by their silence, collaborate with them to enable them to make those decisions.

Before moving on to my conclusion, I want to take seriously Hardy and Clegg's (1996) contention that "a theory of power does not, and cannot, exist other than as an act of power itself" (p. 636). Just as the managers describe the second teacher as someone who has designed his interaction to accomplish his practical purpose of saving his program, so too have I described the managers as designing their report to accomplish their practical purpose of legitimating a decision they want to make. We can thus see my analysis in this chapter as my attempt to accomplish power. I have explicated a perspective on power and the regulation of discourse and advocated an approach to studying these in organizational settings. In doing so, I have asserted my version of power and regulation and have assembled an argument that I hope will make my account count. In the discursive world I have proposed, in which there are no absolute truths, only versions assembled and constructed through language use, my account is but one way of understanding how the world works. I have drawn on the conditions of possibility offered by the academic literature, my own theoretical interests and leanings, my understanding of the organizational imperative to publish so as not to perish, the acceptance of my proposal to write this chapter, my knowledge of the specific conventions for the genre of chapter in an academic book, and myriad other aspects of the social context. I have produced an entry into the academic conversation not because the rules of the genre or the power of the university forced me to. Rather, as a knowledgeable agent, I have oriented to any number of relevant aspects of the context to achieve my practical purpose of persuading readers that my account of power and regulation should be accepted, and that I myself should be regarded as a legitimate voice in the academic conversation. In doing so, I have reproduced some aspects of the social context as the facts of the social world of academia and have resisted or challenged others. My success, or lack thereof, in carrying out my practical purposes is a measure of my ability to accomplish power through language use and interaction and illustrates that communication takes place within a context of perceived power relations and reproduces or challenges those relations.

CONCLUSION

The view of power that I have proposed reveals that the regulation of communicative activity in organizations is not a simple matter of larger forces controlling the activities of individuals in organizations. The rules of particular genres of discourse or the power of particular organization members do not regulate individual communicative activities. Rather participants orient to those rules and power relations and actively reproduce versions of them in their interaction. Power and regulation are collaborative interactional accomplishments. Refusing to see power as something that can be possessed and power relations as an objective reality opens the door to a deeper and more sophisticated understanding of how power is produced and communication is regulated. This perspective on power also underscores that within the limits of the conditions of possibility, power and discourse need not be produced in any particular, predetermined way. The apparently stable facts of social structure do not determine the outcome in any particular interaction.

This perspective on power and regulation has implications for research on communication in workplaces and professions. In particular, the focus of study must be the interplay of the "products of members' reality constructing procedures and the resources from which realities are constructed" (Gubrium & Holstein, 2000, p. 500). This requires attention to both larger social contextual factors and the details of communicative activity. This focus is best achieved through the study of social interaction as it occurs in organizations rather than through more conventional means, such as interviews with organization members about their intentions and motivations in particular communicative events, or think-aloud protocols that ask participants to record their thoughts as they write. These methods produce versions of the organization and the subjects' own participation in the organization that may or may not allow researchers to analyze individuals' activities as both embedded within and constitutive of the organizational context. Studying interaction lets us examine the artful ways in which organization members draw on the conditions of possibility available in organizational settings to accomplish their practical purposes and in doing so reproduce or resist aspects of organizational context.

Understanding power as constructed in interaction also allows us to see why it is that power can slip away so easily. If we understand power as a kind of commodity that people have in varying amounts, it is hard to explain why someone can suddenly have so much less of it. But if we understand it as an interactional accomplishment, we can see that it can never be accomplished once and for all. As managers well know, it must be reaccomplished time and again, every day, in every social interaction. Managers have to assert their version of any given situation over and over, each time producing more power than those who challenge their version. Power is not something that one can ever have; it can only be accomplished through access to interactional resources that allow one to have one's reality claims accepted as the facts of the matter. We can see every communicative interaction as an occasion to reproduce, undermine, or change apparently fixed power relations.

REFERENCES

Anspach, R. (1987). Prognostic conflict in life-and-death decisions: The organization as an ecology of knowledge. *Journal of Health and Social Behavior, 28*, 215-231.

Atkinson, J. R. (1992). Displaying neutrality: Formal aspects of informal court proceedings. In P. Drew & J. Heritage (Eds.), *Talk at work: Interaction in institutional settings* (pp. 199-211). Cambridge, UK: Cambridge University Press.

Atkinson, P. (1988). Ethnomethodology: A critical review. *Annual Review of Sociology, 14*, 441-465.

Bachrach, P., & Baratz, M. (1962). Two faces of power. *American Political Science Review, 56*, 947-952.

Boden, D. (1990). The world as it happens: Ethnomethodology and conversation analysis. In G. Ritzer (Ed.), *The frontiers of social theory today* (pp. 185-213). New York: Columbia University Press.

Boden, D. (1994). *The business of talk: Organizations in action.* Cambridge, MA: Polity Press.

Brandt, D. (1992). The cognitive as the social: An ethnomethodological approach to writing process research. *Written Communication, 9*, 315-355.

Briggs, C. L. (1986). *Learning how to ask: A sociolinguistic appraisal of the role of the interview in social science research.* Cambridge, UK: Cambridge University Press.

Button, G. (1992). Answers as interactional products: Two sequential practices used in job interviews. In P. Drew & J. Heritage (Eds.), *Talk at work: Interaction in institutional settings* (pp. 212-234). Cambridge, UK: Cambridge University Press.

Clayman, S. E. (1988). Displaying neutrality in television news interviews. *Social Problems, 35,* 474-492.

Cyert, R. M., & March, J. G. (1963). *A behavioral theory of the firm.* Englewood Cliffs, NJ: Prentice Hall.

Deetz, S., & Mumby, D. K. (1990). Power, discourse, and the workplace: Reclaiming the critical tradition. In J. A. Anderson (Ed.), *Communication yearbook/13* (pp. 18-47). Thousand Oaks, CA: Sage.

Foucault, M. (1977). *Discipline and punish: The birth of the prison.* Harmondsworth, UK: Penguin.

Foucault, M. (1980). *Power/knowledge: Selected interviews and other writing, 1972-1977.* New York: Pantheon.

French, J. R. P., & Raven, B. H. (1959). The bases of social power. In D. Cartwright (Ed.), *Studies in social power* (pp. 150-167). Ann Arbor, MI: Institute for Social Research.

Garfinkel, H. (1967a). *Studies in ethnomethodology.* Englewood Cliffs, NJ: Prentice Hall.

Garfinkel, H. (1967b). "Good" organizational reasons for "bad" clinic records. In H. Garfinkel (Ed.), *Studies in ethnomethodology* (pp. 186-207). Englewood Cliffs, NJ: Prentice Hall.

Gubrium, J. F., & Buckholdt, D. R. (1979). The production of hard data in human service institutions. *Pacific Sociological Review, 22,* 94-112.

Gubrium, J., & Holstein, J. (2000). Analyzing interpretive practice. In N. K. Denzin & Y. S. Lincoln (Eds.), *Handbook of qualitative research* (2nd ed., pp. 487-508). Thousand Oaks, CA: Sage.

Hardy, C., & Clegg, S. R. (1996). Some dare call it power. In S. R. Clegg, C. Hardy, & W. Nord (Eds.), *Handbook of organization studies* (pp. 622-641). Thousand Oaks, CA: Sage.

Heritage, J. (1987). Ethnomethodology. In A. Giddens & J. H. Turner (Eds.), *Social theory today* (pp. 224-272). Stanford, CA: Stanford University Press.

Hilbert, R. (1990). Ethnomethodology and the micro-macro order. *American Sociological Review, 55,* 794-808.

Hickson, D. J., Hinings, C. R., Lee, C. A., Schneck, R. E., & Pennings, J. M. (1971). A strategic contingencies' theory of intraorganizational power. *Administrative Science Quarterly, 17,* 216-229.

Holstein, J. A. (1988). Court ordered incompetence: Conversational organization in involuntary commitment hearings. *Social Problems, 35,* 458-473.

Holstein, J. A. (1992). Producing people: Descriptive practice in human service work. In G. Miller (Ed.), *Current research on occupations and professions* (Vol. 7, pp. 23-39). Greenwich, CT: JAI Press Ltd.

Holstein, J. A. (1993). *Court-ordered insanity: Interpretive practice and involuntary commitment.* Hawthorne, NY: Aldine de Gruyter.

Holstein J. A., & Gubrium, J. F. (1994). Phenomenology, ethnomethodology, and interpretive practice. In N. K. Denzin & Y. S Lincoln (Eds.), *Handbook of qualitative research* (1st ed., pp. 262-272). Thousand Oaks, CA: Sage.

Hutchby, I. (1996). Power in discourse: The case of arguments on a British talk radio show. *Discourse and Society, 7,* 481-497.

Loseke, D. R. (1989). Creating clients: Social problems work in a shelter for battered women. In J. A. Holstein & G. Miller (Eds.), *Perspectives on social problems* (Vol. 1, pp. 173-193). Greenwich, CT: JAI.

Loseke, D. R. (2000). Lived realities and formula stories of "battered women." In J. R. Gubrium & J. A. Holstein (Eds.), *Institutional selves: Troubled identities in a postmodern world* (pp. 107-126). Oxford: Oxford University Press.

Lukes, S. (1974). *Power: A radical view.* London: Macmillan.

Mehan, H. (1978). Structuring school structure. *Harvard Educational Review, 48,* 32-63.

Miller, G. (1991). *Enforcing the work ethic.* Albany: State University of New York Press.

Miller, G. (1994). Toward ethnographies of institutional discourse. *Journal of Contemporary Ethnography, 23,* 280-306.

Miller, G. (1997). Building bridges: The possibility of analytic dialogue between ethnography, conversation analysis, and Foucault. In D. Silverman (Ed.), *Qualitative research: Theory, method, and practice* (pp. 24-44). London: Sage.

Miller, L. (1993). Claims-making from the underside: Marginalization and social problems analysis. In G. Miller & J. A. Holstein (Eds.), *Constructionist controversies: Issues in social problems theory* (pp. 53-180). Hawthorne, NY: Aldine de Gruyter.

Mumby, D. (2000). Power and politics. In F. M. Jablin & L. L. Putnam (Eds.), *The new handbook of organizational communication: Advances in theory, research, and methods* (pp. 585-623). Thousand Oaks, CA: Sage.

Mumby, D. K., & Stohl, C. (1991). Power and discourse in organization studies: Absence and the dialectic of control. *Discourse and Society, 2,* 313-332.

Reed, M. (1996). Organizational theorizing: A historically contested terrain. In S. R. Clegg, C. Hardy, & W. Nord (Eds.), *Handbook of organization studies* (pp. 31-56). Thousand Oaks, CA: Sage.

Sacks, H. (1992). *Lectures on conversation (1964-1972).* G. Jefferson (Ed.). Oxford: Blackwell.

Salancik, G., & Pfeffer, J. (1977). Who gets power—And how they hold on to it: A strategic contingency model of power. *Organizational dynamics, 5,* 3-21.

Schegloff, E. A. (1991). Some reflections on talk and social structure. In D. Boden & D. H. Zimmerman (Eds.), *Talk and social structure: Studies in ethnomethodology and conversation analysis* (pp. 44-70). Berkeley: University of California Press.

Schneider, B. (2000). Managers as evaluators: Invoking objectivity to achieve objectives. *Journal of Applied Behavioral Science, 36,* 159-173.

Schneider, B. (2001). Constructing knowledge in an organizational setting: The role of interview notes. *Management Communication Quarterly, 15,* 227-255.

Schneider, B. (2002). Theorizing structure and agency in workplace writing: An ethnomethodological approach. *Journal of Business and Technical Writing, 16,* 170-195.

Silverman, D., & Jones, J. (1973). Getting in: The managed accomplishment of "correct" selection outcomes. In J. Child (Ed.), *Man and organization: The search for explanation and social relevance* (pp. 63-106). London: Allen and Unwin.

Simon, H. (1976). *Administrative behavior*. Glencoe, IL: Free Press.

ten Have, P. (1991). Talk and institution: A reconsideration of the "asymmetry" of doctor-patient interaction. In D. Boden & D. H. Zimmerman (Eds.), *Talk and social structure: Studies in ethnomethodology and conversation analysis* (pp. 138-163). Berkeley: University of California Press.

Watson, D. R. (1990). Some features of the elicitation of confessions in murder interrogations. In G. Psathas (Ed.), *Interaction competence* (pp. 263-295). Washington: University Press of America.

Whalen, J., Zimmerman, D. H., & Whalen, M. R. (1988). When words fail: A single case analysis. *Social Problems, 35*, 335-362.

Wilson, T. P. (1991). Social structure and the sequential organization of interaction. In D. Boden & D. H. Zimmerman (Eds.), *Talk and social structure: Studies in ethnomethodology and conversation analysis* (pp. 22-43). Berkeley: University of California Press.

Zimmerman, D. (1969). Record-keeping and the intake process in a public welfare agency. In S. Wheeler (Ed.), *On record: Files and dossiers in American life* (pp. 319-354). Beverly Hills, CA: Sage.

Zimmerman, D. H., & Boden, D. (1991). Structure in action: An introduction. In D. Boden & D. H. Zimmerman (Eds.), *Talk and social structure: Studies in ethnomethodology and conversation analysis* (pp. 3-21). Berkeley: University of California Press.

Zimmerman, D. H., & Pollner, M. (1971). The everyday world as phenomenon. In J. Douglas (Ed.), *Understanding everyday life* (pp. 80-103). London: Routledge and Kegan Paul.

PART III

Critical Research Perspectives

CHAPTER 10

Discourse and Regulation: Critical Text Analysis and Workplace Studies

Brenton Faber

How are forces of change and resistance regulated within organizational contexts? How might researchers better examine, measure, or critique such forces? This chapter argues that by intersecting macro accounts of organizational change with analysis of specific discursive features, researchers can better recognize, examine, and understand regulatory forces within workplace contexts. This chapter examines the discourse used to implement a new software product at a university. Working with specific text features, the chapter articulates the dynamics between technical and social forces that bridge and enhance operations of organizational power. Examining the intersections of macro and micro discursive features can be useful for holding agents accountable for their regulatory actions within changing (and resisting) organizational sites.

In a 1992 essay, Thomas Huckin described what he called "context sensitive text analysis" (p. 84) as a method for combining the linguistic analysis of written texts with an intertextual understanding of the contexts within which such texts are produced and interpreted.[1] Huckin predicted that over the next decade "the

---

[1] Whereas Huckin (1992) used the term *context sensitive text analysis* (p. 84), those from other backgrounds have used differing labels, including discourse analysis, critical discourse analysis, text linguistics, and systemic functional linguistics to describe studies that examine the constitutive relationship between language and social context. In this chapter, I use the wider term *critical text analysis* to include a broad range of practices with the recognition that each school or approach has different emphasis and differently articulated political positions. What draws these practices together is their focus on the constitutive relationship of text and social context.

linguistic analysis of written texts" would become a "major component of composition research" (p. 84) as this emphasis on understanding texts-within-context would help to bridge macro accounts and descriptions of workplace cultures with the micro practices that function within and constitute these sites.

A foundational claim advanced by those practicing critical text analysis continues to be that the processes by which successful communicators produce texts are not isolated, decontextualized, expressivist, or technical activities, but instead are interpretive acts embedded within community participation, social affiliations, and personal choices (Huckin, 1992, p. 85). By using critical text analysis to examine the macro/micro dynamics of organizational regulation, this chapter will show that the complex issues of organizational regulation, like any communicative activity, are social acts and choices that take place as multiple, coordinated, discursive activities. Although many accounts overtly frame and naturalize regulation as a technical procedure or instrumentalist requirement, critical text analysis reminds us that such so-called necessities are still based in social processes that are constructed by ideology, community, societal affiliations, and personal choice.

In other words, critical text analysis provides a method for denaturalizing technologized regulatory processes, challenging such processes for a more transparent accountability of the specific reasons agents provide for their regulatory actions. This chapter will argue that an increased analysis of and attention to the syntactic and pragmatic (micro) features of written texts enables researchers to connect specific regulatory choices to interpersonal, community, and other social (macro) motivations. By establishing these micro/macro connections, researchers may then articulate the coordinated meaning/events[2] inscribed within and against the activities undertaken by regulatory texts.

The chapter is structured in three sections. First, it examines the use of text analysis in workplace research, arguing that, unfortunately, such analysis has been limited to studying topics, key words, and narratives rather than engaging grammatical and pragmatic issues relevant to meaning making. The section notes that there is equally a political as well as a methodological rationale for a greater focus on the full range of how texts construct meaning. Second, the chapter turns to a case study of the implementation of a desktop software program at a small university. Using critical text analysis, the section highlights the ways this regulatory project became unstable and how regulatory agents responded to this instability by turning to social arguments. What was initially defined as a technical project was actually manifested as a social one. The chapter concludes by revisiting the call for closer analysis of regulatory discourse.

---

[2] Texts actively construct and reflect meaning within social situations. The term *meaning/event* is used to refer to this dual process in which a text is both a meaning (reflective of given meanings) and an event (constructing new meaning within a situation).

The purpose of this chapter is not to position detailed critical text analysis as the sole method required for studying issues of regulation. Instead, I claim that examining organizational regulation as a discursive process makes explicit the social issues that are often elided in regulatory discourse. That is not to say that these are the only issues involved in regulatory acts nor that they are sufficient to define the full process of regulation. However, I do offer that social issues are often masked or obscured by arguments asserting technical necessity, procedural efficiency, or other material (nonsocial) reasons for regulatory action. Critical text-based study can therefore act as a reminder of the social implications and objectives of organizational action and a caution against perspectives of organizational processes that elide such key issues.

THE PARTIAL FULFILLMENT OF CRITICAL TEXT ANALYSIS IN WORKPLACE RESEARCH

In order to better recognize the ways specific discourse practices reinforce and construct larger societal and community regulatory practices, it is important to reemphasize the micro/macro dynamic within critical text analysis. To an extent, Huckin's (1992) prediction has been partially fulfilled as linguistic analysis has become an important component of the study of workplace and professional texts. Text-based research has focused on specific activities of workplace writers and the ways their writing construct both formal and informal cultures, relationships, and work processes (Berkenkotter, 2001; Bhatia, 1993; Connell & Galasinski, 1998; Fairclough & Chiapello, 2002; Geisler, Rogers, & Haller, 1998; Henry, 2000; Hyland, 2001; Iedema, 1999; Kelly-Holmes, 1998; Kitalong, 2000; Lemke, 1999; MacDonald, 2002; Schryer, 2000; Segal, 1993; Sullivan, 1997; Swales & Rogers, 1995; Winsor, 2000, 2003; Yeung, 1998; Zachry, 1999). Simultaneously, researchers in complementary fields such as organization studies, education, social psychology, and cultural studies have taken what has been termed a "linguistic turn" in their own research (Alvesson & Kärreman, 2000, p. 136).

As Alvesson and Kärreman (2000) point out, this linguistic turn has been based in the collective research finding that "the proper understanding of societies, social institutions, identities, and even cultures may be viewed as discursively constructed ensembles of texts" (p. 137), or, as Chouliaraki and Fairclough (1999) state more strongly, the "social is built into the grammatical tissue of language" (p. 140). By grounding empirical research about workplaces, organizations, and social groups in specific language practices, including grammatical analysis, researchers have been able to point to key social functions, identity constructions, and processes of active interpretation occurring within these contexts (Chouliaraki & Fairclough, 1999, p. 148).

Perhaps one of the more extensive and influential projects to align with critical text analysis at the macrofunction level has been the study of workplace narratives. By moving away from the concept of master narratives, researchers began looking at the characteristics and the function of local narratives within specific workplaces, individual occupations, and professional practices (Freed, 1993). Perkins and Blyler (1999) identified this as "a narrative turn" (passim) in workplace and professional communication studies. As they wrote, narrative is central to workplace life as "a means of being and of acting" (p. 4), a complex perspective through which agents define and position themselves, and a vital interpretive device for understanding workplace contexts (pp. 4-5). Thus, workplace narratives enable agents to mediate between personal lives and their occupational ones, and they provide a critical means for positioning the self within changing structural contexts of communities at work and home (Faber, 2002). In these ways, narratives have been reported as crucial instruments of interorganizational collaboration (Hardy, Lawrence, & Phillips, 1998), ethical decision making (Dragga, 1997; Sullivan & Martin, 2001), and organizational change (Perkins & Blyler, 1999, p. 24; Faber, 1998, 2002).

Even though ample work has pointed researchers to the micro/macro text dynamics of organizational study, the fulfillment of text-based analysis in workplace and professional communications research has still remained only partial (see Faber, 2003a, p. 421). As Giltrow (1998) has argued, qualitative and ethnographic studies of professional and workplace communication have tended to remain less oriented toward sentence-specific text features and less focused on the syntax, structural characteristics, and pragmatic agency of specific discourse features. When qualitative work has examined texts, the focus has tended to remain on topics or key words rather than at more functional or structural features. Halliday (1994) has responded to this trend by arguing that a critical text analysis that is not based on grammar "is not an analysis at all, but simply a running commentary on a text" (pp. xvi-xvii). His concern is that such an analysis will be too trivial to be meaningful (e.g., the number of words per sentence), or the interpretation itself will miss key features of text cohesion, which identifies how the text and the social context are held together. For Halliday, a full analysis of a workplace text must demonstrate how the semantic (content) and the functional (grammatical, pragmatic) aspects of a text work together to construct the systems of meaning that the text enacts.

Whereas Giltrow's and Halliday's critiques focus on the accuracy and integrity of critical text analysis, Deetz (2003) has recently provided a more critical, epistemological challenge to the linguistic turn in workplace research. He has argued that the widespread adoption and acceptance of the linguistic turn in organizational studies has come at the price of diluting the rationale and the critical perspectives that initiated the turn's original formation. For Deetz, the linguistic turn was a key consequence of social constructivist theory that emphasized "the recognition of the constitutive conditions of experience and the decentering of the human subject as the center or origin of perspective" (p. 422). For social/constructivists, objects of social study were not perceived to exist in nature prior to observation, but were the results of a "constituting activity in relation to the world" (p. 422). Following this recognition, language becomes the focal point for analytical research because it is through language practices that social objects become constituted, gain meaning, and struggle to achieve recognition and power.

However, Deetz (2003) claims that in its most widespread manifestation, the linguistic turn was implemented without a full analysis or appreciation of constitution. As a consequence, a form of text research emerged that simply treated texts as "mirrors of nature" (p. 425). Abandoning a constitutive framework, this "sender-receiver" form of study simply described the text as relaying object. Deetz argues that such research does not provide a critical account of how objects/powers/practices emerge, how they are regulated and challenged, and how they change. Instead, what remains are relativistic, subjective accounts that privilege a personal standpoint or perspective. In other words, in a quest for broad acceptance and utility, various advocates of text analysis traded critical micro/macro dynamics for rudimentary accountings of content or an author's word choice. These accounts are held up at the expense of political accountability and an interrogation of the creation and dissemination of actual cultural and social power (p. 425). Deetz calls for a more robust linguistic turn that better examines the interdiscursive formation of power, the multiple perspectives that negotiate and construct meaning in social contexts, and the development of alternative methods for understanding, explaining, and solving social problems.

A Renewed Epistemology for Critical Text Analysis

As Deetz and Huckin separately have argued, the critical analysis of workplace, organizational, or professional texts is equally a methodological and a sociopolitical project. In this way, Deetz's call for a "serious" engagement with the epistemology of the linguistic turn re-articulates a framework for examining the process and the products of organizational regulation as discursive events. By recognizing the social within the grammatical, the linguistic construction of regulation can be a key focal point for language analysis (van Dijk, 1993, 1998; Chouliaraki & Fairclough, 1999, p. 140). Thus, by examining regulation as discourse, we can view it, in technocratic, market-based, or knowledge-based forms, as social activity, permeated with key issues of political accountability, power relations, and identity construction.

The next section undertakes the task of examining a specific regulatory event as a process that is complicated by the interplay of micro-level discursive actions and macro-level conditions. The section examines the implementation of a new software package at a small university. In summary, my purpose is to provide an analysis of a contentious text in a troubled regulatory process. The analysis will demonstrate how an examination of micro and macro text features can denaturalize the social forces at work mandating the regulation. In this example, I will argue that the e-mail authors are attempting to solidify an increasingly unstable implementation campaign and, by implication, regulatory regime. Although the narrative and semantic content of the e-mail do not overtly suggest that the implementation may be in trouble, closer analysis points to micro features that suggest greater instability. To respond to this instability, the writer turns to social issues, positioning change as natural and normal and asserting that those who do not change are abnormal, uncooperative, and marginalized from the rest of the community.

Thus, the regulatory activity is not based on the merits of the software (technology) itself but on social pressure to conform to the institution's stated goals of normalcy. The analysis demonstrates that although this appears to be a technical (technocratic) context, regulation was instead an overtly social act, attempted through peer pressure, negative identity construction, and personalized *ad hominem* arguments (activities of image construction/power). The analysis also demonstrates how the resistant activities of campus users were integrated within the discourse of regulation to achieve a new construction of normal.

DISCOURSE AND REGULATING INFORMATION TECHNOLOGY: WORKWARE

In July 2002, a small university decided to implement a new information-technology platform. In what was termed a "strategic partnership," the university and a corporate partner agreed on a project that would see the corporation's software, WorkWare (a pseudonym), installed as the only supported desktop software on campus for e-mail (messaging), collaboration, and scheduling. According to a central Web site used to promote the on-campus installation of WorkWare, the software was implemented "to dramatically reduce support overhead caused by the comparatively large number of dissimilar, and incompatible messaging clients and scheduling systems currently being used on campus." WorkWare operates through a common Web page accessible by most Windows-compatible browsers. From this Web page, users log in to access e-mail, scheduling, and collaboration tools. The software includes address books, e-mail tracking, and packages for arranging meetings.[3]

The implementation consisted of two separate but related products. Two new servers were acquired to replace existing e-mail servers. The new server implementation necessitated that all campus users point their e-mail clients to these new servers (akin to plugging into a new box). WorkWare was then positioned as the official software to work with the new servers and the only campus-wide e-mail and collaboration software supported by the campus Office of Information Technology (OIT). Ironically, campus users were never officially *required* to switch to WorkWare. However, OIT was mandated to

---

[3] See Faber, 2003b for an analysis of other e-mails in the implementation.

implement the new corporate partnership, which consisted of both the servers and the WorkWare software.

Chronology of the WorkWare Implementation

The public, written phase of the implementation campaign consisted of seven separate e-mails from July 29 to December 17. E-mail 1 introduced the project as a mandate from the university President, who called WorkWare a "benefit" that came from the university's partnership with this corporation. The e-mail then described the software, outlined the implementation schedule, and articulated several presumed benefits of adopting WorkWare (for analysis of this e-mail, see Faber, 2003b). Subsequent e-mails introduced the implementation schedule (E-mail 2) and announced that the first phase of the implementation had been completed (E-mail 3). During the first three months of the implementation, the e-mails presupposed that all campus users needed to adopt the new software. This was accomplished through key textual silences (Huckin, 2002), which did not mention that users could retain their existing e-mail software. Instead, the e-mails claimed that all users needed to adopt WorkWare.

E-mail 3 admitted that although the campus was aware of the new software, not all faculty and staff had switched to WorkWare. Nevertheless, E-mail 3 confidently claimed that the next two weeks were reserved for switching over those who had not yet adopted WorkWare. E-mail 4 (see Figure 1) was sent to the campus five weeks later. It was the first e-mail directed at those who had not yet switched systems and those who were resisting adopting the new software.

Macro Dynamics:
Power, Duration, Change, Resistance

As theorized above, a regulatory event takes place in the intersection of macro/micro text dynamics. In my examination of the WorkWare campaign, four macroconcepts—power, duration, change, and resistance—emerged as issues relevant to the implementation process and to the methods OIT used to attempt to persuade campus users to adopt WorkWare. Before turning to the microanalysis of e-mail four, I will articulate these four concepts and show how they provide an important frame for the implementation process and my subsequent interpretation of that process.

Power

If social power can be seen as the "self-reflective ability to control an image" (Faber, 2002, p. 143) the WorkWare campaign was about power in two related ways. First, the WorkWare software was a technology that controlled and framed the ways faculty, students, and staff used their communication technologies to represent themselves within public workplace contexts. Thus, the ability and

1. The office of information technology requests that you contact us with your e-mail requirements.

2. If you would like to be migrated to the new WorkWare system, if you would like WorkWare training, or if you would like to discuss your problems or concerns with WorkWare, please contact [representative] at 123-4567.

3. If, however, you plan to remain with your current e-mail client we still need to move your mail service to the new servers [Server 1] and [Server 2], the servers used for WorkWare.

4. This is necessary because the current mail server in use by [the university] is over 4 years old and is starting to show its age, especially with the increase in e-mail traffic over the last several years.

5. The process simply changes the location of where your mail is picked up and delivered.

6. This will require a technician to visit your desktop, but it is painless process [sic], and in most cases, will be totally transparent.

7. Contact [representative] to set up your appointment.

8. If you have specific questions pertaining to this switch, please call the HelpDesk [x1234] and we will have a technician work with you to explain the changes.

9. In order to better serve the campus community we request that you share the reasons you have for not moving to the new system.

10. Please forward your comments to [representative].

Figure 1. E-mail memo 4.

legitimacy to control this software were directly related to the ability and legitimacy to control the ways campus users represented themselves in electronic public spaces. Second, the attempt to convince users to switch their desktop software necessitated the construction of a powerful voice that could control the image of the software, the image of the users, and the image of OIT, which was responsible for instituting the new software. If OIT could construct and manage the various images associated with WorkWare, the implementation campaign would more readily stabilize and permeate the campus.

Duration

Once the transition to WorkWare was complete, OIT needed to establish a system that provided ongoing support and training to WorkWare users. Such support would ensure that they did not switch back to familiar products. Duration refers to stabilized practices across time (Bazerman, 1999, p. 335). It is not enough for a new system to simply displace existing technology, it must establish an enduring set of what Bazerman calls "significant and stable meanings" within discourse systems (p. 335). The process of installing a new software system required that WorkWare be integrated within the campus discourse system, achieving more than a temporary role as a new campus technology.

Change

OIT needed to persuade users to switch their current e-mail, scheduling, and collaboration software to WorkWare, but they could not force them to. They needed to find a way to implement a conscious and acceptable change within this organization. Change can be understood as the reconstitution of a discordant identity (Faber, 2002, pp. 39-40; see also Dutton, Dukerich, & Harquail, 1994). Before change can occur, agents must demonstrate discordance within the existing system and establish a context that is receptive to the new system. Acts of change require the discursive construction of a dysfunctional context and the perception that a new discourse will successfully reconstitute the dysfunctional context.

Resistance

Defined in this case as users' various attempts to destabilize the process (e.g., unwillingness to adopt the software; questions, criticisms, and issues raised about WorkWare; re-adoption of prior software), resistance can be seen as a constituting feature of change. Once OIT began implementing WorkWare, its team needed to address campus resistance to the new software. Without this resistance, OIT could not argue for a robust implementation process, which included training, user forums, support mechanisms, and methods to solicit campus feedback. Resistant practices required that OIT develop a well-supported and persuasive implementation mechanism for engaging campus users within the new discursive context WorkWare established.

These contextual issues are relevant to understanding the situation that created the need for E-mail 4. However, as I argued above, a more detailed understanding of the discursive activity enacted by the e-mail requires a micro-oriented analysis. In what follows, I begin this more detailed analysis at the level of the clause (Halliday & Martin, 1993), which Gee (1999) defines as "any verb and the elements that cluster with it" (p. 98). Cook and Suter (1980) similarly write that a clause is a "construction whose basic characteristic is that it contains, minimally, a subject-predicate relationship" (p. 37).

Working from Halliday's (1994; Halliday & Martin, 1993) typology of meaning (see also Lemke, 1995, p. 41), I have characterized the clausal analysis as (1) *Syntactic:* to identify the organizational function (Lemke, 1995) of language at the grammatical level (what Halliday calls the "textual"); (2) *Semantic:*

to identify the propositional function, meaning how the text constructs the material through its vocabulary and word choice; and (3) *Pragmatic:* to identify the interpersonal and orientational stances of the text.

The significance and degree of these three meanings/functions varies from text to text. In this analysis I present the syntactic analysis first as it details the declining discursive stability of the argument at this stage of the implementation. Next, I present the semantic analysis to detail what the e-mail presents and argues in response to the lack of discursive stability. Lastly, I present the pragmatic analysis to detail the relationships constructed in the e-mail and the orientation the e-mail constructs toward users, those who have not yet changed their software, and those resisting the WorkWare implementation.

Syntactic Analysis:
Text Instability via Relative Clause Structures

Syntactically, there is a high use of relative clauses throughout the e-mail. This high dependence on clausal structures indicates a degree of text instability within the e-mail and within the implementation campaign as a whole. Integrated with the contextual analysis, this instability reflects and constitutes resistance from campus users, a struggle to control the image of WorkWare (power), and acts of regulation as a social process as proposed solutions to the instability.

Research in applied linguistics has connected the use of clauses with the relative stability of context within a text. Halliday and Martin (1993) note that highly claused texts have cohesive structures that place meaning as a process currently underway, reflecting in-process happenings and actions. The authors describe this form of cohesion as a dynamic style. Alternatively, texts with fewer clauses present degrees of abstraction from local accounts to more generalizable, nominalized claims. Halliday and Martin describe these forms of cohesion as a synoptic style, noting that they reflect things, categories, and static structures rather than in-process happenings (pp. 46-49).

Since the dynamic style reports an activity in-process, it represents an event as less stable than the synoptic style which reports a completed, abstracted event. By making this distinction, researchers are not judging the author's level of commitment to the claim nor the relative truth of the claim. Instead, "stability" indicates the degree to which the claim can be said to be completed and abstracted rather than fluid and still in process or development.

For example, in his study of physics research articles published in the *Physical Review*, Vande Kopple (1998, 2002) showed that during the time period he studied, articles used fewer clauses, and the clauses themselves became more lexically dense (more nouns, simple main verbs, bases of infinitives, adverbs, and adjectives per clause) (2002, p. 234). Vande Kopple (1998) claimed that in reducing, over time, their use of clauses, the physics community "displayed different degrees of confidence about the accuracy of scientific instruments, the purity of experimental materials, and even the trustworthiness of experimental results" (p. 184). Vande Kopple correlated the decline in clauses to a shift from the "dynamic style" to the "synoptic style" described by Halliday and Martin (Vande Kopple, 1998, p. 184). This movement from the dynamic to the synoptic style demonstrated a new perspective of physics as a generalizable, stabilized science rather than as a study in flux or a report on physical occurrences and happenings. In this case, the use of fewer and more lexically dense clauses suggested a greater sense of certainty and stability within the discourse and the social community.

In his research, Vande Kopple cited a sentence-to-clause ratio of 1:1.82 for early articles (dynamic) and 1:1.53 for later articles (synoptic). The sentence-to-clause ratio for E-mail 4 (see Figure 1) is 1:2.9, a higher dynamic style than Vande Kopple's data, which suggests an unstable and fluid context. In fact, E-mail 4 has the highest sentence-to-clause ratio of all seven e-mails, even higher than the first e-mail that introduced the software implementation campaign (1:2.4). This suggests that the context for E-mail 4 was even more dynamic and potentially unstable than it was when the implementation was started.

In addition to the high sentence-clause ratio, the discursive instability is also apparent in the functions of each clause and in the layering of prepositional phrases and subject complements throughout the e-mail. For example, sentence two contains three prepositional phrases, sentence three has three prepositional phrases, and sentence four contains four prepositional phrases and three subject complements. These phrases are embedded within the relative clause structure, and they function as modifiers of the main noun and principle verb. In other words, the prepositional phrases and subject complements redefine or reposition the main sentence constituents. Meaning here is literally transient, as the grammar itself resists a stable interpretation.

Semantic Analysis:
Propositional Content Denaturalizes Assumptions

The grammatically indicated instability of E-mail 4 is supported by semantic content in the narrative. Propositionally, this e-mail is important because, for the first time in the implementation, the writer admits that users do not have to use WorkWare. This alternative was not made available in the previous three months of the implementation, and it is not situated specifically but inferred by the statement, "if, however, you plan to remain with your current e-mail client we still need to move your mail service to the new servers . . ." (sentence 3). This concession is significant, as it contradicts the messages that were used initially to implement the software. In other words, it shows readers that the all-or-nothing assertions from previous messages were not accurate.

When integrated with the implementation context, these claims weaken the duration of the implementation project, the "stabilized practice across time," and the claims weaken the ability of OIT to control the public image of WorkWare. The e-mail also provides legitimacy to those who have resisted changing to WorkWare. Having acknowledged that some users may choose not to use WorkWare, the authors face a new problem since they must still find ways to persuade already-moved users to retain WorkWare (duration) and convince resistant users to adopt the new software (change).

Pragmatic Analysis 1: Identity Construction

E-mail 4 uses the metaphor "migration" to identify those who have adopted the new software. Those who do not migrate are left behind and cut off from the main group. Those who have not yet migrated are identified to "have requirements," "problems," and "concerns." In addition, the e-mail infers that those who do not switch to WorkWare are afraid of new technology (it is a painless process) or unintelligent ("have a technician . . . explain the changes"). Those who resist the change for technical, political, or even pedagogical reasons (e.g., Linux, Unix users) are not given an appropriate discursive space or identity. Here, we can see the operation of power, as the writer attempts to control the image of campus users who have not yet adopted WorkWare. This new argument enacts an interpersonal orientation as it marginalizes those who have not yet switched, and it attempts to construct them as uncooperative and abnormal. This new interpersonal orientation is an attempt to gain power and an attempt to regain representational (constructing) authority within the situation.

Pragmatic Analysis II:
Power and Agency Construction

Agency, as one's ability to construct and enact a social identity (Fairclough, 1992, p. 45), is a function of power. To control one's agency is to speak as an empowered subject. Similarly, to control and define another's agency is equally an act of power. The intersection of power and agency is an important issue in the e-mail. The writer transitions agency (and ultimately accountability) for implementing the new software from OIT to individual users. By transitioning agency, the writer is better able to define the image of resistant users. Previous e-mails positioned OIT as the agent of change and users as passive recipients of the new software. However, this e-mail articulates a new relationship between OIT and users. Here, those who have resisted WorkWare appear to be given agency and asked to enact this agency. However, this agency is constituted as agency to adopt WorkWare or agency to be uncooperative and marginalized.

The e-mail begins by requesting that users contact OIT "with your e-mail requirements." The request presupposes that users have requirements and that these requirements should be raised as problematic issues. Three potential e-mail requirements that become potential subject positions (images) for users are then offered in sentence two: (1) those who would like to be migrated; (2) those, already migrated, who would like training on the new system; and (3) those with problems or concerns with WorkWare. Other possible subject positions not included are those who are content with their current system; those who have operating systems that may be incompatible with WorkWare (Linux, Macintosh); and those who have no reason to adopt WorkWare or change their e-mail system. E-mail 4 makes it seem as if users are given agency to identify their e-mail requirements, but the e-mail offers users agency to self-identify only within specific contrasting subject positions—cooperative (normal) or uncooperative (abnormal). The only available choice for those who wish to retain their current software is to identify as abnormal and uncooperative. Moving between the micro and the macro features of discourse, we can see how the textual function responds to the situational instability through an overt act of power (image construction).

The final two sentences request that those who do not wish to adopt WorkWare contact OIT with their reasons "for not moving to the new system." Again, the statement constructs a sense of agency for users while presupposing that users have reasons for not moving. Change is asserted as normal, and those who resist the change must supply reasons for their resistance. The burden of proof is not on OIT to demonstrate why WorkWare is a desirable product. Instead, the burden is on those who wish to retain their current software to demonstrate why they do not want to change.

CONCLUSION: DYSFUNCTIONAL DISCOURSES AND THE RHETORIC OF REGULATION

E-mail 4 concludes by asking those who do not adopt WorkWare to share their reasons for not adopting the new system. Throughout the narrative of the e-mail, change is situated as normal and accepted while resistance to change is situated as abnormal and deviant. Resistance is also used as a rationale for acts of power (image construction). Power is asserted again at the conclusion of the e-mail where persons situated as uncooperative users are asked to defend their reasons for not accepting the change. Yet, this request from OIT is written in the dynamic style (four clauses, two prepositional phrases). Recognizing this adoption of the dynamic style allows a critical reader to recognize that the implementation context remains unstable and indeterminate.

This analysis has examined syntactic (clauses, grammatical cohesion), semantic (narrative meaning), and pragmatic (agency, subject identity) discursive features to provide an example of how one regulatory regime used overt social discourse: peer pressure, negative identity construction, and personalized *ad hominem* arguments to attempt to rescue a destabilized implementation campaign and regulatory program. The lack of stability is evident in the grammatical structure, which strives to construct cohesion while admitting transitional meanings. The argument presented in the text has little to do with WorkWare itself but instead personalizes the implementation, constructing marginal and uncooperative subject positions for those who have not yet migrated. Implementing the software became not a technical issue but an issue of personal identity and conformity with the organization's social norms.

By highlighting the role of the social within a technical process, the chapter demonstrates the social aspects of regulation and the ways regulatory discourse can take place as simultaneously multiple, coordinated discursive actions. Although the WorkWare implementation was construed as a technical procedure (perhaps enriched by the corporatist agreement between the university and the corporate sponsor), the discourse of implementation shifted to assert personal and social claims. This shift in claims, from technical to social, demonstrates a need for critical studies that challenge the accountability of regulatory practices and the reasons various agents provide for regulatory actions. This analysis also seeks a better understanding of technical procedures that claim to have little social impact or relevance. The technical includes the social and cannot be fully disengaged from social implications and determinants.

Texts are a juncture between regulation and agency, the technical and the social, and the organization and society. A key challenge for future text research into regulatory practices is the development of empirical methods that effectively denaturalize acts of power, make power better understood, and hold agents accountable for acts of power. By bridging macrodescriptions of organizational cultures with microanalysis of the discourse of organization, research can better recognize and engage the social and political accountability associated with regulatory practices. The contradictory and socially charged environment of organizations and regulation makes these sites well-suited to such investigation.

REFERENCES

Alvesson, M., & Kärreman, D. (2000). Taking the linguistic turn in organizational research. *Journal of Applied Behavioral Science, 36*, 136-158.

Bazerman, C. (1999). *The languages of Edisons light*. Cambridge, MA: MIT Press.

Berkenkotter, C. (2001). Genre systems at work: DSM-IV and rhetorical recontextualization in psychotherapy paperwork. *Written Communication, 18*, 326-349.

Bhatia, V. K. (1993). *Analyzing genre: Language use in professional settings*. London: Longman.

Chouliaraki, L., & Fairclough, N. (1999). *Discourse in late modernity: Rethinking critical discourse analysis*. Edinburgh: University of Edinburgh Press.

Connell, I., & Galasinski, D. (1998). Academic mission statements: An exercise in negotiation. *Discourse and Society, 9*, 457-479.

Cook, S., & Suter, R. (1980). *The scope of grammar: A study of modern English*. New York: McGraw-Hill.

Deetz, S. (2003). Reclaiming the legacy of the linguistic turn. *Organization, 10*, 421-429.

Dragga, S. (1997). A question of ethics: Lessons from technical communicators on the job. *Technical Communication Quarterly, 6*, 161-178.

Dutton, I. E., Dukerich, J. M., & Harquail, C. V. (1994). Organizational images and member identification. *Administrative Science Quarterly, 39*, 239-263.

Faber, B. (1998). Towards a rhetoric of change: Reconstructing image and narrative in distressed organizations. *Journal of Business and Technical Communication, 12*, 217-237.

Faber, B. (2002). *Community action and organizational change: Image, narrative, identity.* Carbondale: Southern Illinois University Press.

Faber, B. (2003a). Creating rhetorical stability in corporate university discourse: Discourse technologies and change. *Written Communication, 20*, 391-425.

Faber, B. (2003b, October). Technologizing change: Rhetoric of software implementation at a university campus. In D. Novick & S. Jones (Eds.), *Sigdoc 2003: Finding real world solutions for documentation: How theory informs practice and practice informs theory. Proceedings of the 21 annual international conference on documentation* (pp. 171-177). New York: ACM Press.

Fairclough, N. (1992). *Discourse and social change.* London: Polity.

Fairclough, N., & Chiapello, E. (2002). Understanding the new management ideology: A transdisciplinary contribution from critical discourse analysis and new sociology of capitalism. *Discourse and Society, 13*, 185-208.

Freed, R. (1993). Postmodern practice: Perspectives and prospects. In N. R. Blyler & C. Thralls (Eds.), *Professional communication: The social perspective* (pp. 196-214). Newbury Park, CA: Sage.

Gee, J. P. (1999). *An introduction to discourse analysis: Theory and method.* London: Routledge.

Geisler, C., Rogers, E., & Haller, C. (1998). Disciplining discourse: Discourse practice in the affiliated professions of software engineering design. *Written Communication, 15*, 3-24.

Giltrow, J. (1998). Modernizing authority: Management studies and the grammaticalization of controlling interests. *Journal of Technical Writing and Communication, 28*, 265-286.

Halliday, M. A. K. (1994). *An introduction to functional grammar* (2nd ed.). London: Edward Arnold.

Halliday, M. A. K., & Martin, J. R. (1993). *Writing science: Literacy and discursive power.* Pittsburgh, PA: University of Pittsburgh Press.

Hardy, C., Lawrence, T., & Phillips, N. (1998). Talk and action: Conversations and narrative in interorganizational collaboration. In D. Grant, T. Keenoy, & C. Oswick (Eds.), *Discourse and organization* (pp. 65-83). London: Sage.

Henry, J. (2000). *Writing workplace cultures: An archaeology of professional writing.* Carbondale: Southern Illinois University Press.

Huckin, T. (1992). Context-sensitive text analysis. In G. Kirsch & P. Sullivan (Eds.), *Methods and methodology in composition research* (pp. 84-104). Carbondale: Southern Illinois University Press.

Huckin, T. (2002). Textual silence and the discourse of homelessness. *Discourse and Society, 13*, 347-372.

Hyland, K. (2001). Bringing in the reader: Addressee features in academic articles. *Written Communication, 18*, 549-574.

Iedema, R. (1999). Formalizing organizational meaning. *Discourse and Society, 10*, 49-65.

Kelly-Holmes, H. (1998). The discourse of western marketing professionals in central and eastern Europe: Their role in the creation of a context for marketing and advertising messages. *Discourse and Society, 9,* 339-362.

Kitalong, K. (2000). You will technology, magic, and the cultural contexts of technical communication. *Journal of Business and Technical Communication, 14,* 289-314.

Lemke, J. (1995). *Textual politics: Discourse and social dynamics.* London: Taylor and Francis.

Lemke, J. (1999). Discourse and organizational dynamics: Website communication and institutional change. *Discourse and Society, 10,* 21-48.

MacDonald, M. (2002). Pedagogy, pathology and ideology: The production, transmission, and reproduction of medical discourse. *Discourse and Society, 13,* 447-468.

Perkins, J., & Blyler, N. (1999). Taking a narrative turn in professional communication. In J. Perkins & N. Blyler (Eds.), *Narrative and professional communication* (pp. 1-34). Stamford, CT: Ablex.

Schryer, C. (2000). Walking a fine line: Writing negative letters in an insurance company. *Journal of Business and Technical Communication, 14,* 445-497.

Segal, J. (1993). Writing and medicine: Text and context. In R. Spilka (Ed.), *Writing in the workplace: New research perspectives.* Carbondale: Southern Illinois University Press.

Sullivan, D., & Martin, M. (2001). Habit formation and story telling: A theory for guiding ethical action. *Technical Communication Quarterly, 10,* 251-272.

Sullivan, F. (1997). Dysfunctional workers, functional texts: The transformation of work in institutional procedure manuals. *Written Communication, 14,* 313-359.

Swales, J., & Rogers, P. (1995). Discourse and the projection of corporate culture: The mission statement. *Discourse and Society, 6,* 223-242.

Vande Kopple, W. J. (1998). Relative clauses in spectroscopic articles in *The Physical Review,* beginnings and 1980. *Written Communication, 15,* 170-202.

Vande Kopple, W. J. (2002). From the dynamic style to the synoptic style in spectroscopic articles in the *Physical Review*: Beginnings and 1980. *Written Communication, 19,* 227-265.

van Dijk, T. (1993). Editors forward to critical discourse analysis. *Discourse and Society, 4,* 131-132.

van Dijk, T. (1998). *Ideology: A multidisciplinary approach.* Thousand Oaks, CA: Sage.

Winsor, D. (2000). Ordering work: Blue-collar literacy and the political nature of genre. *Written Communication 17*(2), 155-184.

Winsor, D. (2003). *Writing power: Communication in an engineering center.* Albany: SUNY Press.

Yeung, L. (1998). Linguistic forms of consultative management discourse. *Discourse and Society, 9,* 81-101.

Zachry, M. (1999). Management discourse and popular narratives: The myriad plots of total quality management. In J. Perkins & N. Blyler (Eds.), *Narrative and professional communication* (pp. 107-120). Stamford, CT: Ablex.

CHAPTER 11

The Antenarrative Turn in Narrative Studies

David M. Boje

In this chapter, I challenge and problematize the recognized theoretical perspective on narrative in order to pose an alternative that better allows researchers to conceptualize the emergence of new discursive knowledge and power practices. This alternative is something I call antenarratives. Antenarratives are texts that are in the middle and in between what has traditionally been labeled a narrative. They are texts that refuse to attach the beginnings and endings needed to achieve narrative closure. These antenarratives are fragmented, non-linear, incoherent, collective, unplotted, and pre-narrative speculation; they are a bet that a proper narrative can be constituted. The contribution of antenarrative is that it helps to theorize all the non-linear, almost living storytelling that is fragmented, polyphonic and collectively produced—none of which fits into traditional definitions of narrative theory. Looking at the relation of narrative and antenarrative shifts our analytic focus to flows that are not regulated and networks that are emergent but also self-deconstructing, becoming interpenetrated and constantly unraveling and reconstructing storylines.

Narratology, or "the theory and systematic study of narrative" (Currie, 1998, p. 1), is an expansive and active enterprise within which theories of narrative are plural and wonderfully varying. This plurality accommodates perspectives informed by diverse theoretical traditions, including realism, structuralism, and social constructionism. The limitations of each of these perspectives, however, has led poststructuralists (as well as some postmodernists and critical theorists) to deploy the practices of deconstruction and to declare the death of dominant narratology paradigms. For example, deconstructionists have noted that in realist (including pragmatist and formalist) narratology, narrative is a sign system separated from knowledge of the signified; it is a rhetorical device and a contextualist epistemology of historical events unfolding into the present (Boje, 2001, p. 15). In structuralist narratology—which includes the work of American structuralists and French structuralists—narrative is dualized as over story (Culler, 1981, p. 169). That is, narrative is treated as elite to story in that it adds plot and coherence to tidy up fragmented and nonlinear storylines. And, in social constructionist narratology (after Berger & Luckmann, 1966), narrative theory is alleged to exclude political economy, ecology, and ideology in its application to organizational studies.

As deconstructionist critics argue, then, narratives (as they are commonly considered by most scholars) complement modernist conceptualizations of discourse. In these modernist conceptualizations, some overarching principle is introduced to regulate what counts (and, consequently, what does not count) as a narrative. By executing their varied regulatory maneuvers, these scholars ultimately discount or altogether overlook the fragmented and unformed stories that accomplish intriguingly different communicative purposes from those texts that these scholars count as proper narratives.

In this chapter, my intention is to move beyond the duality of narrative and story. This move is necessary, I contend, because current thinking in narrative theory has resulted in a status degradation of story by elite academics who dismiss story as folklore, not a true and integral aspect of narratology itself. To begin, I will briefly illustrate how the impulse to dualize narrative and story affects the work of narrative analysts. I will then move on to presenting a theory of antenarratives, which I propose as a way of accounting for stories within the realm of narratology itself. In the final section of the chapter, I will illustrate the interplay of antenarratives and narratives in the context of recent events at Enron.

THE EXCLUSION OF STORY FROM NARRATIVE ANALYSIS

Narrative analysts seem to replace folk stories with less messy narrative emplotment to create accounts of organizational events that are fictively rational, free of tangled contingency and emergence. Czarniawska (1997), for example, says narrative, or "a *story* consists of a plot comprising causally related episodes that culminate in a solution to a problem" (p. 78). Elsewhere, Czarniawska (1998) clarifies that "for them [stories] to become a narrative, they require a *plot*, that is, some way to bring them into a meaningful whole" (p. 2). A similar definition can be found in Gabriel (2000), who conceptualizes storytelling as narrative "in the narrow sense . . . with simple but resonant plots and characters, involving narrative skill, entailing risk, and aiming to entertain, persuade, and win over" (p. 22). Similarly, Weick (1995) focuses on stories that "gather strands of experience into a plot" or a "good narrative" that provides a "plausible frame for sensemaking," a way of mapping formal coherence on "what is otherwise a flowing soup" (p. 128). Narratives then provide (say narrative theorists) a grander plot, a more emotional charge, a modernist tidying up of stories, whose plots are too rudimentary, missing, or unaesthetic.

Illustrating this point is a study by Czarniawska (1997) in which she collected field notes, documents, and self-reports from her interlocutors. She categorized these using Van Maanen's (1988) typology of stories, serials, and themes. Following this typology, she dualizes serials and themes (which do not have plots) from stories, which have tidy plots of causally related episodes and culminate in a problem solution. Continuing with this perspective, Czarniawska then views stories as the basic building block (or unit) of any narrative.

> In organizations the prescription for a good story is very simple indeed: mix well some random events, several attempts at control, and the corresponding amount of countercontrol, put in a warm place, and wait for results. (p. 79)

Such a perspective on organizations is tied to an inherent assumption that there must always be fully plotted storylines, capable of being told anew as tidy narratives that support power structures. To illustrate the limitations of narrative theory working from this assumption, consider, for example, Czarniawska's (1997) analysis of "A New Budget and Accounting Routine in Big City." In this analysis, she tells the story (actually many stories) of a new city budget and accounting routine to show how narratives "are created and negotiated" (p. 92) out of accounts of local events and intentions. The story bits that are subsumed under the tidy narrative of bureaucratic paradox in her analysis include two quoted anonymous story fragments, which are reproduced in the two paragraphs here.

> Story Fragment 1
> You must understand that municipal accounting today is little else than a game with numbers. Our ambition was to change this state of affairs, to make an annual report actually say something. We wanted to relate it to actual operations and to tell the story of what we wanted to achieve, what has actually been done, and what the prospects are for the future. . . .

> Story Fragment 2
> It is very difficult to abandon old habits—for all those people who deal daily with such questions in a practical context. They may even agree with you in principle, but when it comes to practice, all they see are the problems. They refuse to see the opportunities for us in the change proposed. It's too threatening to their bookkeepers' souls. Some of them have built up all their professional lives around those bits and pieces that must eventually fit elegantly into their places in the final accounts. A proposal that aims at shaking the pieces out of place also means that they will have to lose some of their power and status. (pp. 80-81)

In this story quoting, that Czarniawska turned into narrative, a *paradox* is neatly incorporated into the design of a new accounting procedure, a change that would improve the status quo (p. 99). More formally stated, she poses the "the bureaucratic paradox" as "bureaucracy as a means contradicts its own goal, in the sense that the less humane it becomes, the better (that is, the more smoothly) it functions in the service of humanity" (p. 94). In other words, these storytelling fragments become reinscribed into a simpler narrative emplotment of change and resistance to change, and—through this reinscription—the Swedish bureaucracy blunts an innovation that would interrupt routinized action by translating the innovation into a version of ideas much less threatening to the dominant institutional thought structure (pp. 98-99). The translation and interpretation of local polyphonic stories of events into a single, reductive narrative of paradox is, as Czarniawska (1997) keenly observes, an act of co-optation (p. 96). Counter to this, I would like to resituate narrative's tidy project by looking more closely at the narrative theorist's moves.

As Burke (1945/1952) argues, when "a writer" (in this case Czarniawska), "seems to get great exaltation out of proving, with an air of much relentless-ness, that some philosophic term or other has been used to cover a variety of meanings," (in this case, the term is *paradox*), it is likely because the term "happens to be associated with some cultural or political trend from which the writer would dissociate" herself (p. xiii). Czarniawska's disassociative move is to invoke *paradox* so as to purge storytelling of its localness, and polyphonic mess, making narrative the dominant anecdote to ambiguity in complex organizations. Burke's move is to invoke the pentad elements of act, scene, agent, agency, and purpose (a reinvention of Aristotle's six poetic elements, collapsing rhythm and dialog into agency). If we can follow Burke to an important point here (despite the functionalist nature of his theory), then our task is not to label and dismiss ambiguity; it is to "clarify the *resources* of ambiguity" (p. xiii, italics in original). In this case, the resource of ambiguity in enactment (Burke, 1945/1952, p. 328) is the synecdoche of making simple narratives stand for the whole complex of storytelling dynamics. When this ambiguity is removed, narrative and story are in dialectic relationship, and the transformation is linguistic, as well as dramatic (p. 402).

In her own pragmatist theorizing, Czarniawska substitutes narrative/paradox for multivoiced storytelling to account for the ambiguity of linguistic trans-formation as well as resistance. "Paradoxes," says Czarniawska (1997), "until recently the villains of organizational drama, constitute its dynamics and, even more important, account for its transformation" (p. 97). In Czarniawska's accounting, then, stories and story fragments are simply part of the paradox. However, reframing stories and story fragments as paradoxes that function alongside proper narratives in some sort of narrative/paradox does not adequately account for those stories and story fragments in terms of either what they are or their relationship(s) with narratives. As a way of more adequately understanding stories and their dialectic relationship to narrative, I outline a theory of ante-narratives in the next major section of this chapter.

Narrative theory, as the example taken from Czarniawska's work illustrates, has something important to say about the discursive narration of complex events. Simply put, narrative theory offers a system of representation with rules and practices to produce meaningful statements about a topic. It affords a way of constructing what Foucault (1979, 1980) would call a regulated discourse. Narratives are regulated discourse in that they are constructed through rules of inclusion and exclusion about discursive topics at historical moments. Yet, to me, knowledge about a given topic emerges from more than just the forma-tion of tidy, cohesive plots, after which storytelling dynamics are abandoned to concepts such as paradox. Knowledge also acquires topical relevance and authority within organizations by embodying and conveying various untidy truth claims (or emergent regimes of truth)—be they those of writers or those of the *in situ* characters.

So, while I agree with narrative theorists that narrative is part of the picture of the hastening of the decline of old discourse rules and practices, and that narrative aids the emergence of new discourse by narrating ruptures and radical breaks as acts of paradox or other dramatic emplotment, it seems equally impor-tant to note that there is something more—something that is swept away by narrative closure.

Narratives are a part of the discursive formations that sustain, disrupt, or compose regimes of truth by supplanting story dynamics with reduction. Nar-ratives circulate in multicentered networks of institutional power, but so, too, do lesser-formed storytellings that narratives displace. Narratives are in relation to specific knowledge and power practices and are therefore conveyors of dis-cursive regulation, but so are weaker constructions of story bits. Narratives are self-deconstructing, vulnerable to every new storyline that unweaves and reweaves the hierarchy. And, to me, this is the reconstructive gap that an antenarrative theory can explore.

ANTENARRATIVE THEORY

Antenarrative has a double meaning, emerging from the senses of *ante* being that which is before and also a bet. Used as an adverb, *ante* combined with *narrative* means earlier than narrative. An antenarrative is a gambler's bet that a before-story (prestory) can take flight and disrupt and transform narrative practice. Antenarrative derives its organizing force in emergent storytelling where plots are not possible, or are at least contested and speculative—rich in polyphony and polysemy. Stories are antenarrative when told without the proper plot sequence and mediated coherence preferred by narrative theorists. Antenarra-tives lack the cohesive accomplishment of narratives and do not as yet possess their closure of beginning, middle, and ending. Antenarrative is a nonlinear, fragmented, emergent account of incidents or events, after which narrative comes to add more plot and tighter coherence to the storyline. Antenarratives are abundant, interactive, emergent, and cease to be when they become narrative forms.

My antenarrative focus draws on Derrida's idea of self-deconstruction, a process that occurs as the dynamic weave of differences unravels narrative, with or without the help of the analyst. It also extends to Latour (1996) who argues there is a difference between the linear narrative *diffusion* model (narratives that erupt fully formed in the mind of Zeus) and the nonlinear *whirlwind* model (p. 118). It is this whirlwind model with its rhizomatic ensembles that want to change the world that I call antenarrative. In short, antenarrative is the obverse of the diffusion model of narrative.

Antenarratives are self-deconstructing, fragmenting, emergent, and networking among polyphonic characters; they are not at all static or linear. As I suggested earlier, the crisis of narrative theory in modernity is what to do with the nonlinear antenarrating. How best can we account for the fragmented, polyphonic stories that emerge from everyday situations that are inherently fragmented and defined by simultaneous actions? While telling stories that lack coherence and plot is contrary to modernity, people are always working and living in the middle of collectively mediated antenarrative processes, where few accounts attain narrative closure and fixity.

An antenarrative, therefore, is a bet that a prestory can be told and theatrically performed that will enroll stakeholders in intertextual ways by transforming the world of action into theatrics. At the same time, the antenarratives never quite get there; they unravel as fast as coherence is applied. Antenarrative dynamics include the plurivocal (many voiced), polysemous (rich in multiple interpretations), and dispersed prenarrations that interpenetrate wider social contexts. Antenarrative, therefore, is intertextual: a dialogic conversation among many tellers, listeners, and texts. This antenarrative intertextuality exists within narratives, between narratives, and between antenarratives, fostered by a carnivalesque crowd of authors, readers, performers, directors, editors and spectators all engaged in textual (re)production, (re)distribution, and (re)consumption.

Over the past decade, antenarrative has proven to be a productive concept for scholars interested in little intertextual stories that are not quite cohesive and formed enough to merit being named as narratives. Vickers (2002), for example, looks at how "postmodern antenarratives encourage the possibility that there may be no story to tell, only fragments that may never come together coherently" (pp. 2-3). She combines Heideggerian phenomenology with an antenarrative exploration of multivoiced ways of telling stories, of putting fragments together. Using in-depth interviews of people whose lives were shattered by chronic illness and suffering, Vickers calls attention to that which does not fit into coherence narratives. A study by Barge (2002), takes an antenarrative approach to organizational communication and managerial practice by focusing attention on ways people manage the multivoiced nonlinear character of organizational life.

Barge (2002) says, for example, that antenarrative "requires managers to recognize the multiplicity of stories living and being told in organizations" (p. 7). He gives examples of the managerial practice in the Kensington Consultation Centre in London. I have used antenarrative in my continuing study of Enron (Boje, 2002; Boje & Rosile, 2002), including in a coauthored piece (Boje & Rosile, 2003) in which my coauthor and I look at the dialectic between epic and tragic stories of Enron, arguing that "antenarratives are highly interactive, constituting, and constructing evolving and shifting patterns of prestory connections that reterritorialize an emergent labyrinth that can veer out of collective authorial control" (p. 87).

TOWARD A DIALECTIC THEORY OF NARRATIVE AND ANTENARRATIVE

Extending this earlier work in antenarrative, I want to explore more widely in this chapter how in practice narrative is dialectic to antenarrative. As I will argue, narrative and antenarrative are dialectically related in organizationally situated, regulated discourse. Narrative combined with antenarrative can thus broaden our inquiries into the discourse of organizations, allowing us to examine the multivoiced and emergent ways discursive regulation occurs.

The relationship between narrative and antenarrative can be understood in terms of several important dialectic dimensions. The first dimension is the dialectic between antenarrating and narrating, between emergence and coherence. Secondly, a dialectic exists between divergence into epic antenarrating and the convergence into more tragic (or dramatic) narrative closure. Thirdly, there is a dialectic between antenarratives that are multistranded, multiauthor tellings and the pressure in organizations for collective consensus around official single-author narratives.

An organizational narrative erases a prior way of telling the story, which is to say that each narrative also functions as an antenarrative, demonstrating how one way of telling can erase a prior way of narrating. This is because there is always an inherent pull within narrative to unravel into many-stranded ways of telling and many voices of remembering. Grand narratives are not banished, as Lyotard (1984) would have it, but in practice are opposed and resisted by antenarrating voices. Antenarrative theory problematizes Lyotard's approach to grand narrative by foregrounding the fact that dominant narratives are opposed and shattered by a web of antenarratives. That is, while narrative attempts to camouflage antenarrative fabric and presents a tidier, grand account, situated acts of antenarrating are constantly cracking that façade. A dialectic interplay of grand narration and local antenarration is always at play.

Ricoeur's (1984) emplotment hermeneutics provide a useful way of understanding the link between antenarrating and narrating. In Ricoeur's hermeneutic spiral, antenarrative is the first mimesis, the prestory and the preunderstanding needed to grasp the emergence of plot. The second memesis then is the plot, which required the preunderstanding to make it intelligible. And the third mimesis is the knowledge needed to follow and understand the plot that gets told. In addition to being the first mimesis, antenarrating is also what connects Ricoeur's three mimetic moments, the liminal moves and flows betwixt and between. Ricoeur argues that readers cannot follow a narrative plot (second memesis) through what I am calling its *antenarrative* twists, turns, contingencies, coincidences and dead ends to a forgone conclusion without a great deal of preunderstanding (first memesis) and followability (third memesis). Following this line of thinking, antenarrating and narrating is therefore a dialectic and hermeneutic of time, coherence, and emergence.

In sum, antenarrating in its dialectic relation to narrating lets us focus on how emergence and fragmentation are continually breaking free of closure. To illustrate this point, I will now turn to an examination of the play of narrating and antenarrating at Enron Corporation.

ENRON NARRATING AND ANTENARRATING

Antenarratives have a trajectory in the spectacles that defined events at Enron. In a Deleuzian sense, antenarratives took flight in and through a series of spectacles at that company, yet they also can be used to crack open and transform the public image or faciality of Enron. The faciality of Enron comes through in its press releases, slogans, speeches, reports and Web pages. As I discussed earlier, antenarratives feed on new contexts, they consume contexts, they recontextualize. In Enron's case, antenarratives include not only the blatant theatrical icons of Star Wars, but also plots such as spectacle-turned-into-scandal. Enrollment in galumphing Enron antenarratives took over a decade, but it was unraveling from the very beginning in ways antenarrative theory makes clear.

Table 1 includes a brief chronicle of events related to Enron to help readers recall the unfolding news events in the public narrative. This chronicle, which focuses on the dramatic shifts in corporate strategy that Enron executives gave out to the public, helps to set the stage before I begin showing the more rhizomatic network of antenarratives playing into and against this chronology.

Within each of these corporate shifts there is a multiplicity of strategic plots. In Table 2, we (Boje & Rosile, 2002) outline 10 plots that different commentators use to make sense of Enron.

The 10 plots in Table 2 represent the dialectic of tragic and epic looks at Enron. Plots 1 and 2, for example, focus the tragedy on a few players, thereby protecting layers of the political and economic system from more epic plots of 3 through 9. Plot 10 says everyone is to blame, which is to say, no one is because everyone does it.

In the antenarrative approach that I am advocating here, our approach to making sense of these contradictory plots is to shift the focus of analysis from

Table 1. Enron Chronology 1985–2002

- 1985 – Enron sets out to become the premier natural gas pipeline in North America (Enron is the hero of deregulation in every presidential administration's New Economy).

- 1990 – Enron changes its plot to become the world's first natural gas major (Enron is the hero of deregulated Free Economy market ideological frames).

- 1995 – Enron changes plot to become the world's leading energy company (Enron becomes the hero of Virtual Corporations, an ideological frame that is about downsizing employees while providing maximum perks to full-time executives and their staff).

- 2001 – Another plot change, to become the world's leading company (Enron is the superhero of the Bonfire of the Vanities, a world conqueror of American hypercompetitive global capitalism).

- Jan. 2002 – A new plot, to emerge from bankruptcy as a viable, albeit smaller company (Enron, after bankruptcy is re-plotted as the villain, responsible for the *domino effect* of collapse of Arthur Andersen, and the string of expose megaspectacle scandals, such as WorldCom).

- Aug. 2002 – for Enron executives, the plot becomes to keep from serving time in jail

---

"what's the story here?" to questions of "why and how did this particular story emerge to dominate the stage?" In this sense, antenarrative theory (Boje, 2001) is closely tied to the work of Kristeva (1980) and Bakhtin (1981) wherein each text has an intertextual trajectory that is historical and social (see also Boje, 2002). And antenarrative theory relates to Fairclough's (1992) approach to critical discourse analysis, which advances the idea that the intertextual trajectory is embedded in a hegemonic struggle.

Before I turn to an examination of the interplay of narratives and antenarratives at Enron, then, it will be helpful to introduce two conceptual metaphors that will help make sense of the admittedly confusing events at Enron—events that became confusing by the design of those involved. These two concepts which are borrowed from the work of Deleuze and Guattari (1987), are trajectories and rhizomes.

Trajectories

Antenarrative trajectories take flight as long as there is context left to transform (Deleuze & Guattari, 1987). This is the case because antenarrative is about ontological ways of being in the world. That is, antenarrative is not sense making (Weick, 1995); it is world making (e.g., antenarratives feed on new contexts, they consume contexts, they recontextualize). In terms of the events at Enron,

Table 2. Enron Whodunit Plots.

Plot 1. It was Andrew Fastow and his lieutenants.	Plot 6. It all relates to Afghanistan and the pipelines of ExxonMobil, Texaco, Unocal, BP Amoco, and Enron.
Plot 2. It was Fastow's bosses: Skilling and Lay.	Plot 7. It is a remake of the Seven Sisters from 1911.
Plot 3. It was the Domino Effect—ripple effects of Enron caused other corporations and executives to be exposed.	Plot 8. It was a failure of the system of Western Capitalism (its checks and balances).
Plot 4. It is the Whitehouse; this is Enrongate.	Plot 9. It was the Business College; they taught the gentleman crooks everything they needed to defraud.
Plot 5. It was all three Presidents (Bush Sr., Clinton, and Bush Jr.).	Plot 10. It was the greed and hubris of everyone.

through incessant antenarrating, bits and pieces of real events—along with Star Wars and Jurassic metaphors—were inserted in the executive's dialogs as they made blanket assurances of increasing profitability. Thus, antenarratives flirted between fiction and reality in ways that seduced spectators and stakeholders into willingly suspending disbelief. These antenarrative trajectories at Enron can be productively thought of as strands that take rhizomatic paths.

Rhizomes

A rhizome, as defined by Deleuze and Guattari (1987), is a network of sub-terranean trajectories, including root/stem strands, radices, bulbs, and tubers (pp. 6-7). The alleged fraud at Enron—its off-the-balance-sheet partnerships—is an example of a rhizomatic network. In this network, the raptor partnerships burrowed beneath the surface images of Enron from 1991 (with Cactus) on through 1997 (with Chewco, LJM1, LJM2) and deeper still with the secretive Southampton Place and RADR partnerships. The partnership strands fused and separated, stopped and restarted, encountering and absorbing blocks in rhizomatic fashion. The rhizome of fraud assumed many diverse partner-ship forms while sustaining a faciality or public image of responsible risk management.

The rhizomatic clusters of antenarratives shown in Figure 1 depict the inter-textual ties that bound these clusters together at different times. As this figure

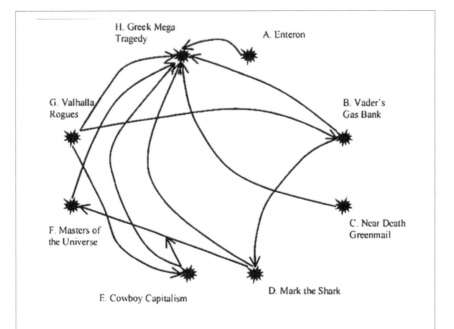

H. Greek Mega
Tragedy

A. Enteron

G. Valhalla
Rogues

B. Vader's
Gas Bank

C. Near Death
Greenmail

F. Masters of
the Universe

E. Cowboy Capitalism

D. Mark the Shark

Key to Figure 1:

A:H Renaming game recommences after collapse of Enron.

B:H Offshore accounts, fake records, and Arthur Andersen of B repeat in H. **B:D** Skilling's 10-year war to oust Rebecca Mark ("the Shark").

C:H As in C, shareholders in H seek control and liquidation of Enron.

D:H Oppressive tales of villagers in India do not play on center stage until collapse. **D:F** Mark's global capitalism plays into Masters of the Universe.

E:(D:F) Cowboy Capitalism combines with Mark the Shark globe hopping strategy.

E:H Houston galas play & morph into White House scandals.

F:H Bonfire of the Vanities interpenetrates with White house, Congress, and UK elite appointments to Enron boards and committees.

G:B Valhalla Rogues resurfaces in Gas Bank.

G:E Valhalla Rogues merges with Cowboy Capitalism.

G:H Off-balance-sheet accounts and offshore banking & Arthur Andersen repeat from G to H.

H Each of the antenarrative clusters reemerges in postcollapse inquiries.

Figure 1. Examples of intertext of antenarrative trajectories.

suggests, the antenarratives of one cluster migrate and interpenetrate with those of other clusters.

The key at the bottom of Figure 1 gives examples of antenarratives that move in between the clusters listed in Table 1. Figure 1 suggests many of the pieces of "H" (Greek Mega Tragedy) of the Enron collapse were present in antenarrative clusters, such as "G" (Valhalla Rogues), the oil traders of 1985–1987 who brought about an SEC investigation, Andersen Audits, offshore accounts, creative accounting processes, and scandal. To take another example, pre and post to the "A" (Enteron) cluster, George Bush Jr.'s Spectrum 7 oil company had Enron partnerships (1985 and 1986); in other words George Bush did not just begin Enteron-relations with corporate and executive contributions to his gubernatorial or presidential campaigns. I would like to expand on one example of antenarrative trajectory, "Gas Bank." Keep in mind, though, that each of the antenarrative clusters in Figure 1 contributes to rhizomatic relations between the trajectories.

GAS BANK AS AN EXAMPLE OF ANTENARRATIVE TRAJECTORY

At Enron, the Gas Bank project lasted from 1989 to 2001 and proved to be more than a collaboration of many public/private institutions, Wall Street analysts, Harvard and Stanford MBAs recruited to be traders, and politicians from countries around the world. The origin of the Gas Bank idea is disputed (and becomes complicated as the idea morphs into the trading floor concept), but whomever the author, the dialog and emplotment went like this: Enron would be an intermediary between buyers and sellers of natural gas, exploiting the spread between the buying and selling price. In this Gas Bank antenarrative, "gas producers" were "depositors" in the "commercial bank" and the "consumers" were the "borrowers"; "Enron" was the "bank" that "pooled the deposits" (i.e., the supply commitments) to fund long-term (15-years or more) commitment to gas buyers (the borrowers). What happened with the Gas Bank idea is also unsettled. In one version, when Skilling presented his Gas Bank antenarrative to the assembled Enron Board and executives, the idea was soundly rejected. However, in yet another version, Kenneth Lay himself decided to ignore their advice and give Skilling a chance to make it work. Others say it was Richard Kinder, Enron's president (and not Kenneth Lay), who asked Skilling to join the company to run the new Gas Bank adventure (Barnes, Barnett, & Schmitt, 2002). Most storytellers agree that on June 29, 1990 Skilling left McKinsey to join Enron. Skilling described Enron's Gas Bank strategy as, "get in early, push to open markets, position ourselves to compete, and compete hard when the opening comes" (Kaminski & Martin, 2001).

In 1989, Skilling (or Bennett) also came up with a "Gas Swap" strategy, to remedy failed negotiations between Enron and a Louisiana aluminum company. When Enron could not physically transport gas from its own facilities and make the Louisiana deal profitable (Kaminski & Martin, 2001), the Gas Swap strategy was developed that "called for the customer to buy gas locally, paying a floating price, and simultaneously purchase a swap from Enron in which Enron would pay the producer's floating rates and the producer would pay Enron a fixed rate" (Kaminski & Martin, 2001, p. 44). The Gas Swap was a metascript fix to the Gas Bank antenarrative; as in a commercial bank model, the Gas Swap would be equivalent to a deposit guarantee system. There were glitches to the plot, however. For example, the initial Gas Bank plan did not persuade gas producers to sell Enron their reserves, so Enron decided to offer money upfront to entice gas producers to deliver gas, later on, at the preagreed price. The problem was where to get the money.

Enter Andrew Fastow, a finance whiz kid from Continental Bank in Chicago, who moves to Houston (which made his wife much happier, since Houston is the home of her prominent family). In 1991, Fastow came up with a way to get the money, via a partnership strategy called "Cactus." Cactus partnerships took in money from banks and lent it to "energy producers in return for a portion of their existing gas reserves" (Barnes, Barnett, & Schmitt, 2002). As with the Gas Bank antenarrative, the Cactus rescripting has disputed beginnings. What is known, however, is that Fastow would become instrumental in the hundreds of off-the-balance-sheet partnerships that would become the Achilles heel of Enron by 2001 (see Figure 1). Fastow had a reputation as a mean-spirited manager; he was Vader's lieutenant and would bear down on subordinates. The other side of Fastow's character was known among the Houston social circle set wherein he was a benefactor to the arts (along with his wife). An important antenarrative is forming at this phase of the concentrated spectacle formation: financial wizards can keep the Gas Bank alive and stop the kind of rogue trading that the Valhalla high-stakes gamblers pioneered in 1985 (see antenarrative cluster G below).

> "If you ask an outsider what industry Enron is in they will say energy. If you ask an insider they will tell you that we are in the risk management business. We provide certainty of delivery and certainty of price." (Andrew Fastow, CFO, Enron Corp., cited in Kaminski & Martin, 2001, p. 44)

To keep the Gas Bank going, Skilling and Fastow implemented a "Market Forward Price Curve" method, based upon Enron's unique knowledge of price information (from gas depositors and customers). Each day, every commodity-trading desk posts a single forward price curve, calculated directly from gas market prices (and used to predict the future price of gas or some other commodity).

There were four significant changes to Enron in 1991 when Skilling and Fastow developed Gas Bank into the precursor to the Energy Bank and Trading Floors. First, the bureaucratic organizational structure inherited from the Enteron

(HNG/InterNorth) merger had to be reengineered; consequently, the company was reduced from 15 layers of management to 4, thereby moving decision authority closer to people who possessed the knowledge. Second, Skilling changed the company's human-resource-management performance-review policy to implement his infamous rank-and-yank system (the goal was to remove 20% of the company's lowest performers each year). Third, management and trading talent were recruited, primarily from Harvard and Stanford business schools' top graduates (luring graduates with $20,000 signing bonuses, $80 K salaries, and annual bonuses of up to 100%). Fourth, the risk-management strategy for gas trades was changed from "experience, intuition, and estimates of potential losses" to a one-unit-sensitivity-risk analysis (for example, one basis-point movement in interest rates, or one-dollar move in the price of crude oil), and finally to a more sophisticated value-at-risk (VaR) approach at the portfolio level (Kaminski & Martin, 2001). This philosophy was combined with recruiting only those business school graduates who were high-risk takers. As we shall see (Antenarrative Clusters E, F, and H) a macho, risk-taking Enron corporate culture developed in the mix of events at Enron: declining energy-market fortunes by 2000, the rank-and-yank performance system, the CIA/FBI e-mail monitoring by the late 1990s, Skilling's Machiavellian strategies, and Fastow's aggressive personality. And, despite the rating by *Fortune* as a best place to work in America, Enron became backstage one of the most oppressive places to work in America.

Gas Bank Morphs

Enron's Gas Bank morphed into deals such as "JEDI LP." JEDI is Enron-speak for "Joint Energy Development Investments." JEDI began in 1993 as a partnership of Enron and the California Public Employees Retirement System (Calpers) to invest in natural gas projects and assets. For its part, Calpers invested $250 million in the partnership, but in 1997 Enron bought back its stake in JEDI from CalPERS for $383 million and immediately sold it to another limited partnership called Chewco, named after the Chewbacca character in *Star Wars*. Chewco was run by a former Enron executive. Then came other partnerships called LJM1 and LJM2 (both run by Mr. Fastow). On October 24, Fastow ($30 million richer) was replaced as chief financial officer and put on leave. On November 8, 2001, Enron decided Chewco, LJM1, and LJM2 should have been consolidated on their books since 1997, thereby reducing prior reported net income by $586 million. At this point, the Securities and Exchange Commission began to investigate. Enron eventually wrote down shareholder equity by $1.2 billion when it closed out its relationships with the LJM partnerships. On top of these, there were also Kenobe Inc and Obi-1 Holdings, which were off-the-balance sheet partnerships that helped to hide losses of $2 billion in the water and telecommunications businesses, another $2 billion lost in a Brazilian utility investment, and an additional $1 billion lost on an electricity-generating plant in India.

These events at Enron provide the basis for now turning more explicitly to the antenarrative versions, the competing prestories, and the contested/unresolved histories, to see what patterns appear. There are emerging concentrated spectacles (e.g., Enteron and Gas Bank), which have yet to marry with more diffuse spectacles (in the next section), and then into more integrated spectacles. There are also early, tamer versions of megaspectacle scandals such as Fastow's Cactus and Skilling's Gas Bank that try to resist being #G, "Valhalla rogue oil traders," but, as we shall see, the offshore accounts and accounting practices of the rogues return again and again to haunt Enron. Finally, in these antenarratives, there is stagecraft, showmanship, front and backstage plots and counterplots, and the assembling of the cast of characters for a set of scripts that is opposed by the builders of pipelines and utility buildings.

Work by Bougen and Young (2000) in rhizomatic studies of auditing and bank fraud is useful for understanding the rhizomatic networks at Enron. Enron's partnerships were assemblages of "heterogeneous collections of people becoming connected in different ways and at different times to" commit fraud (Bougen & Young, 2000, p. 2). A familiar refrain can be applied to both the bank frauds that Bougen and Young exam and those at Enron: "How could this have happened?" How could it have happened when there were so many regulators, auditors, government agencies, and administrators who are supposed to protect investors and employees from fraud? It was largely possible because the rhizomatic lines of partnerships kept on the move, intersecting with one another (Deleuze & Guattari, 1987, p. 203). The strands would connect and disconnect, form and dissolve, then reassemble. It is this constant movement that makes fraud so hard to trace; it just will not stand still. The partnership characters keep changing; the pattern will not remain stable. Consequently, the intersecting strands that get blocked by inquiry find ways to move around the rules and the gaze of scrutinizing parties. From 1890 to 1892, forty-one U.S. national banks failed. Twenty-six failures were the result of fraud. As at Enron, we know something happened. We know the players involved. And we can trace the money movements, but these movements could not be traced by the average Enron investor nor by the average employee.

In the congressional hearings, the Powers (2002) report, and the Houston federal case, efforts are made to freeze the antenarrative assemblage and to get a look at the fraud. But fraud passes undetected at the time when it goes down. Fraud, in fact, was part of Enron since its beginning in 1985, but it kept moving, circulating, assembling. The cast of characters changed, but the character fraud kept playing at the wheel of fortune. And fraud made money for lots of players. Many, it seems, did not want to really detect fraud. They did not want to closely follow the rhizomatic lines to detect fraud; the strands were there, but few were complaining, and none followed up.

Rhizomatic Assemblages of Antenarratives

Antenarrative trajectories form rhizomatic assemblages that can ensnare fraud. It works both ways. The assemblages will reach back to the memories of the Seven Sisters and see the Seventh Sister is Enron. This is the Theater of Capitalism, a dialectic battle between fraud rhizomes and detective rhizomes. It is like the battle on the Internet between virus and virus detection. Without fraud detection, fraud would become endemic to Capitalism. We would never be rid of it. The impersonal rhizomatic and secretive partnerships schemes veil the interests of the perpetrators. It is a detective's task to trace the rhizomatic antenarrative trajectory patterns.

Lay and Skilling and Mark gave capitalism a personal and romantic charm that investors and regulators and politicians could identify with. Rooting out the tubers is just one way to capture much prolonged public attention. When this happens, there will be a brief purge, then back to business as usual. Capitalism will once again put on its magic mask and use the strange hypnotic power of spectacle to delude the "spectator" (Fetter, 1931, p. 4). Once again we shall be as Shakespeare's character Titania in *Midsummer Night's Dream* and fondle the ass's head. The courts will glimpse momentarily the grotesque and burlesque face of capitalism, then be intoxicated by the spectacle and see only romantic surfaces.

My dialectic theory is that Enron begins in an antenarrative rhizome process and ends up in megascandals packaged as coherence narratives with plots to entertain and reeducate us. Yet elements of the narrative refuse to cohere and the forces of antenarrative unravel the romantic plot of Enron as the New Economy role-model corporation that has transformed itself from brick and mortar utility to virtual corporation. Enron's Gas Bank, along with its emerging off-the-balance-sheet special-purpose entities, realizes and then derealizes wealth and reputation for Enron executives. Characters begin to disenroll in the Enron narrative, and antenarrative counterforces disintegrate the romantic plot while Congress and various agencies rescript more tragic narrative plots. There are many comedic and satirical antenarrative unravelings. On the surface, there is a narrative of dramatic changes to accounting rules, executive insider-trading rules, and increased oversight from the SEC, GAO, and a hoard of Congressional oversight committees. Yet, beneath that surface erupts the antenarratives that suggest that narratives of containment are a thin veneer, as capitalism itself seems to veer out of the orbit of human and institutional control.

IMPLICATIONS OF ANTENARRATIVE THEORY

Antenarrative theory holds several important implications for studies of communicative practices in organizations. First, those using narrative methods should no longer ignore antenarrative dynamics. It is important to not only compare narratives but also to see how antenarratives form, reform, and transform in intertextual ways. In the case of Enron, scandal, strange accounting practices, and political influence were ingredients from the beginning.

Second, analyses that refer to a unitary universal narrative (e.g., "The story of Enron is the story of unmitigated pride and arrogance," says Jeffery Pfeffer, as cited in Pearlstein & Behr, 2001, p. A01) should be seen as reductionistic taglines. In my view, they miss the morphing of antenarratives (such as the Gas Bank) and their changing intertextual relationships through complex rhizomatic practices.

Third, fixing narrative terms and pentadic ratios at a single point in time misses how ratios shift across antenarrative trajectories. For example, a Burkean-type of plot ratio can become a shifting ratio over time. There are intertextuality and rhizomatic relationships that must be traced, not just in ratios but also in the storytelling soup from which such narrative taglines are applied.

Fourth, it is important to look at the emergence of the networking of antenarratives in future studies. Antenarratives are self-organizing fragments that seem to cling to other fragments, and form interesting relationships. This is a call for more intertextual research, tracing the way one implicates another in time and in context.

Fifth, each antenarrative conveys an ideological framing by the analyst, which is quickly opposed by counterframes of other analysts (e.g., Table 2 plots). For example, the ideological framing of Enron as the superstar of Free-Market capitalism is opposed by the ideology of Enron and its global pawns where deregulated power costs more and is less efficient, and where Enron is a frequent violator of human rights around the globe. And these ideologies are opposed by one that says that Enron is just greed and hubris or just the fault of auditing fees (that pay a little) mixed with consultancy fees (that pay a lot). The more grotesque frame of critical theory suggests that Enron is the result of predatory and hyper-competitive practices of global capitalism; a more romantic frame argues that the Enron and Arthur Andersen spotlight turned the state's attention on ways to repair 401k plans, close loop holes in annual report notes, change the incentives of executives in reformations of stock sharing laws, and changes to political contributions that are expected to put a cap on corporate and PAC giving. Whether this is an actual repair to the checks and balances of late-modern capitalism or an exercise to restore consumer and investor confidence in business and the state, time will tell. It is likewise important for scholars who want to trace the antenarrative roots of Enron's collapse in 2001 to go back to its beginning in the history of corporation and industry in ways that are rhizomatic and intertextual.

POSTSTORY

In sum, everyone wants a simple story to make sense of complex dynamics, but this is not always helpful. In the case of Enron, the eight antenarrative clusters are contextualizing and recontextualizing processes of change. The antenarratives recruit a cast of characters and constitute one side of the dialectic of story and narrative.

In tracing the antenarrative versions, the competing pre-stories, and the contested, unresolved histories, we can see more clearly what patterns appear. Antenarrative and narrative in dialectic relation makes a contribution to inquiry by exploring the gaps and excesses excluded in traditional narratology.

REFERENCES

Bakhtin, M. M. (1981). Discourse in the novel. In M. Holquist (Ed.), *The dialogic imagination: Four essays by M. M. Bakhtin* (C. Emerson & M. Holquist, Trans.). Austin: University of Texas Press.

Barge, K. J. (2002). *Antenarrative and managerial practice.* Working Paper, University of Georgia.

Barnes, J. E., Barnett, M., & Schmitt, C. S. (2002). How a titan came undone. *U.S. News and World Report.* March 18. On line at http://www.laydoff.com/articles/usnwr_3_18_02.html.

Berger, P., & Luckmann, T. (1966). *The social construction of reality.* New York: Doubleday.

Boje, D. M. (2001). *Narrative methods for organizational and communication research.* London: Sage.

Boje, D. M. (2002). *Critical dramaturgical analysis of Enron antenarratives and metatheatre.* Plenary presentation to 5th International Conference on Organizational Discourse: From Micro-Utterances to Macro-Inferences, Wednesday 24th–Friday 26th July (London).

Boje, D. M., & Rosile, G. A. (2002). Enron whodunit? *Ephemera, 2*(4), 315-327. Online journal at http://users.wbs.warwick.ac.uk/ephemera/ephemeraweb/journal/2-4/2-4bojeandrosile.pdf.

Boje, D. M., & Rosile, G. A. (2003). Life imitates art: Enron's epic and tragic narration. *Management Communication Quarterly, 17,* 85-125.

Bougen, P. D., & Young, J. J. (2000). Organizing and regulating as rhizomatic lines: Bank fraud and auditing. *Organization, 7,* 403-426.

Burke, K. (1945/1952). *A grammar of motives.* New York: Prentice Hall. (First printing 1945; 1952 edition being quoted.)

Culler, J. (1981). *The pursuit of signs: Semiotics, literature, deoconstruction.* Ithaca, NY: Cornell University Press.

Currie, M. (1998). *Postmodern narrative theory.* New York: St. Martin's Press.

Czarniawska, B. (1997). *Narrating the organization: Dramas of institutional identity.* Chicago, IL: University of Chicago Press.

Czarniawska, B. (1998). *A narrative approach to organizational studies.* Qualitative Research Methods Series Volume 43. Thousand Oaks, CA: Sage Publications, Inc.

Deleuze, G., & Guattari, F. (1987). *A thousand plateaus: Capitalism and schizophrenia* (B. Massumi, Trans.). Minneapolis: University of Minnesota Press.

Fairclough, N. (1992). *Discourse and social change.* Blackwell, Oxford.

Fetter, F. A. (1931). *The masquerade of monopoly.* (First edition 1931, NY: Sentry Press. Reprinted 1971 Augustus M. Kelley Publishers, NY. Quotes are from the 1971 reprinted edition.)

Foucault, M. (1979). *Discipline and punish: The birth of the prison* (Translated from the French by Alan Sheridan). New York: Vintage Books.

Foucault, M. (1980). *Power/knowledge*. New York: Pantheon.

Gabriel, Y. (2000). *Storytelling in organizations: Facts, fictions, and fantasies*. Oxford, NY: Oxford University Press.

Kaminski, V., & Martin, J. (2001). Transforming Enron Corporation: The value of active management. *Journal of Applied Corporate Finance, 13*(17), 40-49. Available online at: http://finance.baylor.edu/gis/enron.pdf.

Kristeva, J. (1980). *Desire in language: A semiotic approach to literature and art* L. S. Roudiez (Ed.), (T. Gora, A. Jardine, & L. S. Roudiez, Trans.), New York: Columbia University Press.

Latour, B. (1996). *Aramis, or the love of technology* (C. Porter, Trans.). Cambridge, MA: Harvard University Press.

Lyotard, J. F. (1984). *The postmodern condition* (1979 in French edition and 1984 English edition) G. Bennington & B. Massumi, Trans.). Minneapolis: University of Minnesota Press.

Pearlstein, S., & Behr, P. (2001). At Enron, the fall came quickly. *Washington Post.* Sunday, December 2nd; Page A01. On line at http://www.washingtonpost.com/ac2/wp-dyn/A44063-2001Dec1.

Powers, W., Jr. (2002, February 1). Report of Investigation by the Special Investigative Committee of the Board of Directors of Enron Corp.

Ricoeur, P. (1984). *Time and narrative* (Vol. 1), (K. McLaughlin & D. Pellauer, Trans.). Chicago, IL: University of Chicago Press.

Van Maanen, J. (1988). *Tales of the field*. Chicago: University of Chicago Press.

Vickers, M. H. (2002). *Illness, work and organization: Postmodernism and antenarratives for the reinstatement of voice*. Working paper, University of Western Sydney.

Weick, K. E. (1995). *Sensemaking in organizations*. Thousand Oaks, CA: Sage.

CHAPTER 12

Hearing Discourse

Robert P. Gephart, Jr.

This chapter presents a critical/interpretive approach to the analysis of regulated communication at public hearings and inquiries conducted by government regulatory agencies. The critical/interpretive approach integrates three distinct perspectives on discourse analysis to provide more comprehensive insights into regulated communication than are provided using any single perspective. First, narrative/rhetorical analysis is used to examine the substance and form of narratives and stories of regulation. Second, ethnomethodology is used to analyze the process of sense-making that underlies regulatory communication and makes it meaningful. Third, Habermas' critical theory of communication is used to understand the political context and implications of regulated communication. All three aspects of the critical/interpretive approach are illustrated with analyses of data constituted by testimony of key persons involved in a formal hearing into a fatal pipeline fire.

This chapter presents a critical/interpretive approach that can be used to understand how communication is regulated during public hearings and inquiries conducted by government regulatory agencies. The critical/interpretive approach also provides insights into how regulated discourse at public hearings legitimates key social institutions.

Public hearings and inquiries can be defined, according to Boyer (1960), as meetings organized by government agencies to enable public participation in policymaking and planning (p. 285). I am interested in public hearings and inquiries since they are vitally important events in what Beck (1992) refers to as contemporary risk society (p. 19). Public hearings and inquiries are one of a limited number of ways for the public to directly participate in government (public) policymaking (Boyer, 1960). In public hearings and inquiries, different groups compete for relief from risks that have accumulated during the production of wealth (Beck, 1992). Further, public inquiries and hearings are settings where the regulation of communication by government agencies is publicly visible. Visibility emerges because government agencies explicitly use public inquiries and hearings to review the effectiveness of formal regulations, to identify and rectify problems in regulations or regulated behavior, and to assess and mitigate the risks that wealth production holds for society. Hence, inquiries are key settings where government agencies actively seek to mediate among the interests of different groups. Effective conduct of inquiries and hearings is important to maintenance of governmental and institutional legitimacy because people assess the legitimacy of key institutions by examining the ability of institutions to conduct effective inquiries and to protect the public from harm. The present chapter does not distinguish inquiries from hearings, given the similarities that Salter and Slaco (1981) have established between the two.

I propose in this chapter that the regulation of hearing discourse and its implications for institutional legitimation can be usefully investigated through a critical/interpretive approach to discourse analysis that combines three perspectives: rhetorical/narrative analysis, ethnomethodology, and Habermasian critical theory. Each of these perspectives focuses on different aspects of communication and communication contexts, and each offers unique insights into specific aspects of regulated communication. The first perspective, the rhetorical/narrative, is grounded in anthropological studies of narrative. This perspective uses concepts from literary theory and literary criticism to examine narratives and stories told by members about their experiences of regulated institutional settings. Narratives and stories form the substance of much regulated communication; hence, narrative/rhetorical analysis addresses the *substantive* dimension of regulated communication and the form it takes. Rhetorical analysis complements narrative analysis by showing how selective construction of storytelling influences or regulates understanding and meaning.

The second perspective, ethnomethodology, is based on interpretive sociology. It examines how sense-making practices are used during verbal interaction to provide stable bases for ongoing interaction. Ethnomethodology thus investigates how the stable features of society are produced and accomplished. From an ethnomethodological point of view, culture consists not in the substance of events and phenomena but in the processes through which members of society interpret phenomena. Ethnomethodology therefore addresses the *process* dimension of regulated communication—the deep-level interpretive processes that regulate communication and construct culture.

The third perspective, Habermas' critical theory of communication (1973, 1979), is grounded in critical social theory. As Kemp (1985) explains, Habermasian critical theory focuses on the formal properties of ideal-type speech acts that form the basis for a rationally grounded social consensus in democratic society (p. 188). Habermasian critical theory addresses how micro-level communication influences macrosocial structures and how micro- and macro- features of social order interact to sustain or undermine institutional legitimacy and the general conditions needed for democratic governance. Habermasian critical theory thus addresses the *political contexts* of speech acts and their institutional implications.

The critical/interpretive approach that I am proposing integrates these three perspectives to provide a richly textured analysis of hearing discourse that accounts for more complexity in regulated discourse than does each perspective in isolation. With the integrated approach, narratives and stories of crises and hearings into crises are described in detail and analyzed rhetorically to uncover important features of the substance of communication. Then, general discourse at hearings is analysed to understand how sense-making practices enact proc-esses that produce coherent discussions of events. Finally, hearing discourse is examined in terms of idealized standards of valid communication to understand the political contexts and institutional implications of this discourse. With the integrated approach, a key objective is to understand how hearing discourse impacts the legitimacy of key social institutions. I argue that regulation of hearing discourse distorts communication as a means to privilege the interests of estab-lished institutions and thus to protect their legitimacy.

In this chapter, I provide a general overview of important features of hearings. Next, I discuss the critical/interpretive approach and explain how it provides a more complete and holistic framework for understanding the regulation of hearing discourse and its relationship to institutional power than any of the three individual perspectives. Drawing upon data I gathered from a public hearing involving a fatal oil pipeline fire that occurred in western Canada in 1985, I then illustrate each of the three perspectives, showing how they illuminate hearing discourse when applied separately versus when they are integrated via the critical/interpretive approach. The conclusion discusses how the study of hearing discourse contributes to understanding regulated communication in institutions and organizations, facilitating a cultural turn in communication studies.

PUBLIC HEARINGS AND INQUIRIES: PARTICIPATION IN POLICYMAKING

According to Boyer (1960), the three means for public participation in public policymaking are conferences, advisory committees, and public hearings. A public inquiry or hearing is an investigation conducted by a body, board, or commission legally mandated to assess an issue and allow public participation. An inquiry tribunal or board is accountable to the mandate that governmental agencies provide for the inquiry. The tribunal reports to the government and its recommendations are considered by government agencies before being implemented.

Hearings differ from conferences and advisory committees. Hearings, Boyer (1960) explains, are publicly announced in advance, and any interested person is allowed to attend and testify (p. 285). While hearings are required in some situations, they are sometimes an agency's optional choice. Formal hearings, which are regulated in a manner similar to legal proceedings in court, use such procedures as testimony under oath, transcription of proceedings, cross-examination of witnesses, use of exhibits, legal counsel, and rules of evidence. Hearings, as Boyer points out, can also be informal, resembling the town hall meeting (p. 286). Hearings and inquiries are common in Canada (Slater & Slaco, 1981) and Great Britain (Kemp, 1985), and according to Checkoway (as cited in Cole & Caputo, 1984), they are now "among the most traditional methods for citizen participation in America" (p. 406). Public hearings have been held to investigate many topics, including oil and gas development proposals (Keeling, 2001; Nikiforuk, 2001), attacks on children in a hospital in the United Kingdom (Brown, 2000, p. 51), the slide of a coal mine colliery tip at Aberfan (Turner, 1976, 1978) and the space shuttle Challenger accident (Vaughan, 1996).

Public hearings and inquiries share several important characteristics. First, they are a major site for political discourse in the public sphere (Habermas, 1989; Kemp, 1985)—the domain of liberal democracy where people can participate in discussions with government agencies in an effort to solve important social problems. The public sphere emphasizes openness, impartiality, and rationality in decision making. For Kemp, government policies and decisions gain legitimacy when they emerge from an open, democratic inquiry process. Inquiries, Kemp argues, can justify government decisions in advance by demonstrating that decisions are outcomes of public inquiry processes. This justification occurs implicitly because public participation makes decisions appear more rational and less political than otherwise (p. 183).

A second characteristic of public inquiries is that, despite their assumedly open and public nature, inquiries are forms of regulated communication, underscoring, in Kemp's view, the practices and policies in bourgeois society that restrict public access to the public sphere (p. 181). Such restrictions include governmental regulations, such as laws and organizational policies that constrain the behavior of organizations and the persons directly and indirectly involved in them. Inquiry boards and policies also regulate the actual inquiry participants, speakers, topics, and discussions. Inquiries thus regulate and limit public involvement, with openness, impartiality, and rationality in decision making transformed to disguise the particularistic interests served by inquiries. As Wynne (1980) has argued, these limitations on public involvement keep conflict latent but create additional legitimacy challenges, since publics use inquiries to assess institutions that conduct inquiries and control technology (p. 186).

A third characteristic of public hearings and inquiries is that they are important ceremonials in contemporary culture. Ceremonials, according to Trice and Beyer (1993), amalgamate "discrete cultural forms into an integrated public performance" (p. 109). Ceremonials have technical, practical, and expressive outcomes and a sacred quality (Trice & Beyer, 1993). They are elaborate social dramas with defined roles, they require preplanning, and they involve collective activities with audiences. Gephart's study (1992) shows, for example, that inquiries involve assembling representatives of legitimating and critical institutions as knowledgeable witnesses to the events in question (p. 133). Other studies (Turner, 1976; Gephart, 1993; Brown, 2000) use native accounts of witnesses to analyze inquiry board reports for their policy recommendations and accounts of events under scrutiny. Inquiries, as Trice and Beyer (1993) emphasize, are thus rich in cultural forms and meaning.

Finally, public hearings emerge during the cultural adjustment stage of accidents (Turner, 1976) and social problems. Incidents such as technological disasters can challenge institutional legitimacy by showing that regulatory institutions cannot protect the public from risk and harm. If government regulatory agencies fail to address these problems, they may face legitimacy challenges and loss of public support. A tension therefore emerges between the pursuit of governmental interests and public interests. To address this tension, government regulatory agencies control inquiry conduct and regulate testimony, witnesses, and discourse to create (1) the sense that key institutions enact important democratic values including impartiality, rationality, and fairness that are necessary for institutional legitimacy in liberal democracies; and (2) the sense that key social institutions are capable of protecting the public from harm. Institutions that fail to address social problems may lose legitimacy and public support. Inquiries can thus be characterized, in Emerson's terms, as societal "last resorts" (1981, p. 1) that solicit public input on important issues when prior remedies have failed to resolve problems.

In this chapter, I build on the argument that by regulating discourse, inquiries develop acceptable interpretations of social problems to legitimate key institutions (Brown, 2000; Gephart, 1992, 1993; Gephart & Pitter, 1993). These regulations, which produce distorted communication, limit social conflict that could potentially undermine the legitimacy of social institutions. This chapter targets methodological and analytical tools that can advance insights and understanding concerning this regulation of discourse and communication.

DISCOURSE AT HEARINGS

Discursive views of organization presume linguistic patterns are central to examining organizational life (Putnam & Fairhurst, 2001). Discourse is defined as written- and spoken-language use (Fairclough, 1992, p. 62) that establishes the syntax used to create meanings and comprehension (Riddington, 1990, p. 189).

Past research on the discourse of public inquiries (e.g., Gephart, 1992, 1993, 1997) has uncovered three important dimensions of regulated communication. First, regulated communication has a *substantive* dimension: content, which includes the narratives and stories used to represent events and communicate experience during an inquiry. Narration is defined here as the process of providing first-person accounts of experiences, with narrative as one such account. A story, according to Boje (2001), is a discrete aspect of narration that highlights specific people and the events they experience and provides a plot, bringing people and experiences together in a meaningful whole (p. 2). Narratives and stories can frame and control our interpretation of experience by emphasizing particular features of the world and key characters or persons and by deemphasizing or omitting other features. Thus, it is important to examine narratives and stories of regulation and also the regulations referred to in the narratives and stories to understand how narratives and stories shape our understanding of events. Narrative analysis examines the substance of peoples' stories and how they are, in Riessman's words, "put together" (1993, p. 2) as well as the cultural resources that stories draw upon.

Rhetorical analysis complements narrative analysis. It is directed at analyzing how stories are constructed to persuade readers of the authenticity of experiences and events (Riessman, 1993). A narrative/rhetorical perspective that combines the two modes of analysis is well-equipped to analyze the substance and explicit features of stories and storytelling practices in hearing discourse as well as rhetorical features that shape interpretation.

The second dimension of regulated communication is the *process* dimension that provides for the sensibility of discourse. Face-to-face conversational discourse unfolds over time and across sequential utterances. Such communication is produced and assessed by participants in terms of sensibility. At points where sensibility is a problem, communication breaks down with subsequent interaction directed at repairing breaches of sensibility (Gephart, 1993, p. 1470). Because discourse is controlled or regulated by these sense-making practices that support or undermine the sensibility of particular statements and claims made during discursive interaction, it is important to examine sense-making processes. Ethnomethodology, developed by Garfinkel (1967) and defined by Heap (1975) as the "science of sensemaking" (p. 107) is well-suited to investigate this process dimension of hearing discourse as regulated communication. Ethnomethodology analyzes the sense-making practices of social interaction in actual situations to understand how these practices, conceived as the process of sense-making, construct a sense of shared meaning.

The third important dimension of regulated discourse is its *political context* and implications for legitimacy. Discourse provides bases for reinforcing or undermining the legitimacy of regulatory institutions based on whether the statements in discourse are conceived by participants as valid or invalid forms of communication. Discourse can be controlled by regulatory institutions that seek to reproduce institutional legitimacy through speech acts that assert the validity of claims. This regulation is a separate dimension from storytelling insofar as stories can be based on valid or invalid statements and still be "good stories." Validity and legitimacy of regulatory communication also differ from sense-making since one can make sensible statements that are invalid and senseless statements that are true. Habermasian critical theory (Habermas, 1973, 1979) is well-suited to analyze the validity of speech acts and communication. This perspective uses idealized standards of valid communication to assess whether and how speech acts are free from or regulated by institutional constraints. And it relates the validity of speech acts to the production of legitimacy by showing how communication can be distorted so as to shape interpretations and enhance legitimacy, even where communication was not free from institutional constraints.

THE CRITICAL/INTERPRETIVE APPROACH: THREE PERSPECTIVES

The critical/interpretive approach to regulated discourse in hearings and inquiries is designed to account for the multidimensions of regulation by integrating the narrative/rhetorical analysis of stories, ethnomethodological analysis of sense making, and Habermas-based analysis of speech acts and legitimacy. As such, the critical/interpretive approach provides a rich framework (Gephart, 1992) for addressing how inquiry discourse transforms preliminary interpretations of events into institutionally acceptable accounts consistent with institutional logics that legitimate state regulatory practices.

To illustrate the critical/interpretive approach, the following sections further explicate each of the three perspectives and then demonstrate their integration.

The Narrative/Rhetorical Perspective

A narrative-rhetorical perspective focuses on analyzing the substantive and regulatory effects of stories. Such analysis is basic to the critical/interpretive approach because it targets how narratives shape and reflect understanding, including how individuals and groups understand and interpret events (Boje, 2001; Brody, 1981; Czarniawska, 1998; Riddington, 1982, 1988, 1990).

Advocates of a narrative/rhetorical perspective challenge more traditional views of narrative in which stories are conceived as neutral, realist descriptions of events and actions and in which language is viewed as a mere conduit—or technical device—for transmitting meaning. Analysis informed by a narrative/ rhetorical perspective emphasizes instead the constitutive power of stories and language, targeting narratives as rhetorically constructed artifacts. A narrative/ rhetorical perspective is thus designed to provide insight into the communicative strategies that construct and regulate organizational discourse as well as how individuals experience and interpret this discourse.

Although this perspective provides no single template for conducting analysis—it involves creative work—analysis typically examines an informant's story and how it is put together. According to Riessman (1993), such analysis includes the linguistic and cultural resources a narrative draws on and how it establishes authenticity or credibility with others (p. 2). Riessman points to the structural properties of narratives as especially important features to examine. Following the work of Labov, Riessman argues that a fully formed narrative includes the abstract or summary of the substance of narrative; the orientation of the narrative including time, place, situation, and participants; complicating actions; an evaluation of the actions and the attitude of the narrator; the resolution of the narrative (what happened); and the coda of the narrative (p. 18). Analysis of these and other rhetorical strategies can reveal, says Riessman, how people order their experiences and interpret important events and actions in their lives (p. 2). Such analysis can also, as Brown (2000) points out, reveal the authorial strategies that construct a consistent text, absolve key people, and legitimate professions while avoiding the legitimacy of the state (p. 50). For example, in Brown's analysis of the Allitt inquiry report on attacks on children in a hospital in the United Kingdom, he shows inquiry reports are contrived rhetorical products created to persuade readers to accept a contestable interpretation of events (Brown, 2000). Inquiry reports, he concludes, support the legitimacy of social institutions and extend the hegemony of prevailing ideologies (p. 48).

As an initial step in the critical/interpretive approach to analysis of inquiry discourse, narrative/rhetorical analysis offers a unique focus on the substance and form of stories and on the rhetorical practices of text and construction of meaning that frame and influence interpretation. The usefulness of a narrative/rhetorical perspective can be demonstrated by examining data from an inquiry into a pipeline fire involving highly flammable natural gas liquids (NGLs) (Gephart, 1993). Initially, I offer a realist telling of the story of this fire so as to orient the reader to the events and characters that were involved. Then I provide an excerpt from narrative data and analyze it from a narrative/rhetorical perspective.

The Pipeline Fire: A Realist Story

The pipeline fire under investigation occurred in western Canada in February 1985, during efforts to repair a leak of NGLs in a pipeline. The pipeline maintenance crew was notified about the leak in the afternoon, on a cold winter day with temperatures approximately $-5°F$. The assistant district manager took the call and informed the district manager and also the pipeline maintenance crew manager about the leak. Other crew members were then informed about the leak. The foreman, assistant district manager, and crew members drove to the leak site in a farmer's field some 100 miles from the crew's headquarters. The foreman and a crew member inspected the leak area and established the approximate location of the leak. By 1400 hours the assistant manager and the foreman agreed they faced a major leak of NGLs, a highly volatile substance that can be ignited with a spark. Flaring or controlled ignition is the prescribed procedure for control of such leaks and is undertaken to eliminate unplanned ignition of gas clouds during repair efforts.

Discussions were held between the district manager who remained at headquarters and the assistant district manager and pipeline maintenance foreman who were at the leak site. In the discussions, these personnel debated the wisdom of flaring the leak. By 1730 hours it was too late to flare the leak since it was now dark at the site, and flaring would be dangerous due to problems in monitoring the gas cloud. Repair equipment, including a lighting plant used to illuminate night work, was then driven to near the leak. The lighting plant failed to start and was jump-started, only to stall. At 2020 hours two crew members went to shut off the engines of their vehicles. Suddenly a flame several football fields in size erupted, and the pipeline maintenance foreman and four crew members were engulfed in flames. Other crew members and newly arrived personnel provided aid to the injured workers who were transported to the hospital by ambulance. The pipeline fire burned for three days. The foreman and one of the crew members died two weeks later. Three other workers who received serious burns were hospitalized for several months. An inquiry conducted by a major federal agency was held into the events soon after the accident.

A Regulatory Narrative

During the inquiry, an important concern was the practice of flaring, or voluntary ignition of the pipeline leak of NGL. The leak was not flared, and the subsequent fire seriously burned five workers, two of whom subsequently died. At the hearing, the assistant district manager and others narrated their experiences with the leak and fire. During their testimony, they referred to the company policy manual that stated: "On a major NGL leak the best course of action may be to ignite it. Once ignited the danger of a growing and drifting explosive vapour cloud is eliminated" (Gephart, 1993, p. 1488). The district manager with authority for repair of the leak testified, "I said I will discuss it again [flaring] with Don. But I, personally, didn't want to flare. There is no hiding that . . . There was never, 'no, we won't flare' from me. I hope you don't get the impression it was a closed book. . . . It was never—I expressed my reluctance but I never, in any conversations, said we won't flare" (Gephart, 1993, p. 1492).

Narrative/Rhetorical Analysis

Analysis of the district manager's involvement in the flaring decision reveals several important features of narrative that relate to how people order and interpret experience, and how they attempt to persuade others their interpretations are valid and credible. Using concepts from the narrative/rhetorical perspective discussed above, we can analyze the segment of narrative extracted from the official proceedings to uncover rhetorical properties of the District Manager's statements. The segment provided here is a partial telling of the story about the decision not to flare, and the story is repeated in different form(s) at numerous points in the hearing proceedings.

First, the company policy suggesting that one should ignite an NGL leak provides a potential master narrative to prescribe how effective pipeline maintenance crew members should and presumably normally do handle NGL leaks by flaring them. This master narrative is an important linguistic and cultural resource used by company personnel and government agencies to understand the cause of the unplanned fire. The narrative seeks to establish the credibility of the district manager by claiming that the district manager did not say "no" to flaring; rather, he was merely expressing a view based in the dangers of flaring that made him reluctant to endorse flaring. He uses the analogy that it was not a "closed book" to highlight the lack of decision and also refers to his anticipation of further discussion with Don, his superior, as evidence that no final decision was made by him.

Second, the narrative shows how the form of a story can lead to particular interpretations of experiences and events. The narrative is a partial and not fully formed narrative, and the question-and-answer format that is enforced in the hearing limits the district manager's responses to answers to questions that others pose. The story is oriented to what the district manager was doing at the time of the leak, when he was off site, and how he interacted with others. The story highlights the agency of the manager through statements that construct him ("I") as a central figure in the story. The district manager's narration of his decision not to flare the leak provides a complicating action in the broader story of the leak. The evaluation of actions and attitudes of the narrator is another important formal feature of narratives that has rhetorical implications. In this case, although the district manager provided a narrative that sought to rhetorically distance him from the decision, he emerged as a potentially villainous figure, since he was formally responsible for the decision and clearly did not support a decision to flare. The manager's refusal to accept responsibility for the decision and his acceptance of a flawed view (e.g., "There is no hiding that") show his motives were problematic and he accepted that there were problems with his behavior. This explains the highly problematic resolution of the events in the general story—two people died and three were seriously burned presumably because the wrong decision was made under the authority of the district manager, who failed to follow company policies and sought to avoid responsibility for the decision. Thus by focusing on narratives and stories as important substantive aspects of regulatory discourse, narrative/rhetorical analysis shows how different stories and contradictions reveal alternatives to the main story and provide for a different understanding of events recounted in stories.

A narrative/rhetorical perspective is directly applicable to the analysis of stories and narratives produced during conversation. In general, according to Labov and Waletzky, it is appropriate for analysis of oral, first-person accounts of regulatory experiences that take the form of "natural narrative" (as cited in Riessman, 1993, p. 69). It is also appropriate where the substance and form of narratives and stories are the focus for analysis. One may need to adapt narrative/ rhetorical methods for analysis of written materials such as texts (Riessman, 1993). Further, a narrative/rhetorical perspective is not useful for studies of regulated communication involving large numbers of subjects whose identities are not known (Riessman, 1993). It is not useful for fast and simple research on regulated discourse since it requires extensive, fine-grained analysis. It is also not useful where easy and unobstructed insights into the roles of regulation in subjects' lives are desired, since narratives are complex and many details must be considered in analysis. And it is not a suitable technique where one conceives language as a transparent medium. Further, a narrative/rhetorical perspective does not address the complete discourse, including conversations that occur at hearings; it is not well-equipped to address the full range of processes that produce sensible stories and discourse; and it does not necessarily address the political contexts or implications of regulatory storytelling and narrative. The next approach, ethnomethodology, is well-equipped to more fully examine the broader discourse processes that assemble sensible stories and discourse.

Ethnomethodological Perspective

Ethnomethodology provides a second perspective for understanding how communication is regulated in hearing discourse. Conceived as the science of sense making, ethnomethodology investigates, in Lynch's (1993) words, the "genealogical relationships between social practices and accounts of those practices" (p. 1). Because ethnomethodology assumes that sense-making methods are necessary for any socially organized behavior (Coulon, 1995, p. 16), it is concerned with discovering the underlying sense-making processes people use to produce acceptable behavior in a variety of situations (Feldman, 1995, p. 8). The sense-making methods examined by ethnomethodology occur primarily in conversation, but they can be examined in other forms of communication including texts and documents (Bogen & Lynch, 1989; Leiter, 1980; Lynch & Bogen, 1996; Turner, 1974).

Ethnomethodologists look for processes by which people make sense of inter-actions and institutions (Feldman, 1995, p. 4); that is, how sense is made, not what sense is made. Ethnomethodological methods, as outlined by Coulon, include careful observation and analysis of actual behavior in natural settings to uncover the processes by which social members interpret reality and use language as a resource to build a "reasonable" world (pp. 16-17).

Several concepts from Garfinkel (1967) illustrate ethnomethodology's domain and process focus. One concept is indexicality, which refers to the natural incompleteness of words. Ethnomethodology assumes that all symbolic forms have a fringe of incompleteness that disappears only when they are performed. To fill this gap, words and utterances draw meaning from the contexts in which they are used. Thus a word can be analyzed only with reference to its situation of use, and no absolute meanings are independent of such situations. According to Coulon (1995) indexical expressions constitute discourse itself (pp. 19-20). Another core concept is reflexivity, which refers to the self-constituting character of social settings and descriptions of those settings. Ethnomethodology assumes that social practices of communication describe social frameworks and constitute frameworks at the same time. Thus public hearings produce descriptions of regulatory actions and also implement verbal and other control over interaction. Related to reflexivity is the concept of accountability. Ethnomethodology focuses on the accountability of the social world as it is described and made reportable by social members. A final concept is cultural membership; ethnomethodology addresses the person as a cultural member. Members are persons with an ensemble of processes, methods, activities, and the know-how to make sense of the surrounding world in a way that can potentially be corroborated by other members. A member, Coulon explains, exhibits mastery of the institutional language used by a group and a natural social competence that allows one to be recognized and accepted (p. 27).

Central to ethnomethodology is analysis of the sense-making practices that underlie institutional language and that construct and regulate a sense of shared or intersubjective meaning in institutional discourse. Sense-making practices, which groups use to co-construct meanings, also produce regulation and control by requiring that institutionally appropriate general themes or issues are addressed in conversation and communication, particularly when one responds to statements by other parties to the interaction. Ethnomethodology thus provides a means to investigate the discursive practices of regulation and control on an utterance-by-utterance basis. To do so, ethnomethodology targets two opposed situations in communicative interaction: (1) what Feldman (1995) identifies as breakdowns and (2) situations where breakdowns are nearly impossible because norms are so well-internalized (p. 4). In the latter, the constructed nature of sense-making practices may be invisible to members because those practices are normalized as common sense. During breakdowns, however, sense-making practices are disrupted or breached. In such situations, according to Leiter (1980), meaning begins to disintegrate, and members engage in repair practices that use sense-making practices to restore a sense of shared meaning (p. 217).

Four important sense-making practices addressed by ethnomethodology are discussed here (Leiter, 1980, following Garfinkel, 1967; Cicourel, 1973; Gephart, 1993, 2004). First, people assume meaning is shared if they can produce and sustain a reciprocity of perspectives whereby individual participants assume that if they exchanged places, they would experience the other's view of the world. The reciprocity of perspectives regulates discourse by requiring that all parties to a conversation produce statements that accord with others' perspectives or to make statements that others can at least understand. Perspectives that can readily be taken or understood by a wide range of other persons tend to be worldviews that are common in society and that are institutionally acceptable and legitimate.

Thus the reciprocity of perspectives regulates discourse by privileging institutionally common or dominant views of the world that others can understand and share. Uncommon or unusual perspectives cannot commonly be adopted. Second, members construct and use normal forms of words and terms and expect others to do so. The practice of using and expecting normal forms regulates discourse by encouraging and indeed requiring participants to employ institutionally acceptable and common terms when communicating. Third, members sustain interaction by using an etcetera procedure, which assumes what others are saying (a) will be filled in by the hearer and (b) will be clarified later. The etcetera sense-making practice regulates discourse by influencing the pace of interaction. It permits pauses by speakers and limits interruptions by hearers by requiring parties to conversation to allow speakers to continue making statements even when they are not fully comprehensible, and to wait until later for comprehension to emerge. Fourth, members employ terms as indexical expressions that will be comprehended by the hearer using implicit background knowledge the hearer possesses, including knowledge of the speaker's identity; relevant features of the speaker's biography, purpose and intent; the setting in which the interaction occurs; and the relationship between speakers and hearers (Leiter, 1980, p. 174). This regulates discourse by imposing additional frameworks of interpretation, such as the institutional context of the interaction.

The analysis of sense-making practices is important to the critical/interpretive approach I am proposing because it provides a means to understand how discourse is regulated at the microlevel of utterances in conversation. The analysis of how sense-making practices are used in the regulation of discourse is demonstrated by analysis of testimony provided during an inquiry into the pipeline fire (Gephart, 1992, 1993) discussed above. The testimony involved questions posed by legal counsel for the corporation and answers given by the district manager who was allocated primary responsibility for the accident. The testimony addresses flaring the pipeline leak of natural gas liquids.

> **Question 1:** Do you remember what other reasons [for not flaring] you mentioned to Wayne S.?
> **Answer 1:** Well, I was very aware that the night was coming on, and we certainly didn't want to flare at night. And we really didn't want to babysit a big fire at night. I agreed with him that that was very valid.
> **Question 2:** But on the other side of the question, did you give any reason, any other reasons, besides the Milepost 75 fire, as to why you did not want to flare at that time?"
> **Answer 2**: "Well, the biggest reason is that everything, I felt, was under control. We had . . . we were going to do a normal stopple job."

This example can be used to show how ethnomethodological concepts can be used to analyze and understand how sense-making practices regulate discourse during inquiries. First, the questions by the legal counsel call for the district manager to construct a reciprocity of perspectives between himself and the legal counsel concerning reasons for not flaring the leak. Construction of a reciprocity of perspectives regulates meanings and interactions by confining them to shared realities, and thus the initiation of a reciprocity of perspectives sets up the expectation that the district manager will evidence an institutionally common and sensible view concerning flaring of leaks. However, the district manager shows his lack of shared understanding and his unique view on the issue of flaring, and therefore breaches the emerging expectation of reciprocity.

Second, flaring is a normal form response or procedure for pipeline leaks of natural gas liquids since it incinerates flammable and dangerous gases. It is a legitimate, professionally warranted, and institutionally appropriate response. Thus the questions pose flaring as a normal response to leaks and they construct the Milepost 75 fire (a fatal fire that occurred previously) as a reason the manager did not want to flare. Flaring is not advisable for crude oil leaks where the likelihood of accidental ignition is low. A fire can burn for a lengthy time and delay pipeline repairs. This problem occurred when the Milepost 75 fire ignited under the district manager's supervision several years prior to the present leak. During the Milepost 75 fire, two workers received fatal burns.

Third, the questions create an incipient etcetera principle since they propose or offer the district manager an opportunity to elaborate his understanding of flaring and the Milepost 75 fire. In response, the manager breaches the expectation of reciprocity and reveals his lack of shared understanding. He considered the leak to be under control and claimed stoppling was the appropriate normal form response that would solve the problem without flaring. The manager thus produced a unique interpretation of events that contradicted the sense making of others at the hearing. He offered the Milepost 75 fire as a reason to avoid flaring. This further undermined his credibility since the Milepost 75 fire was an unplanned and uncontrollable ignition of crude oil that should not be relevant to the current situation involving a leak of natural gas liquids. The manager failed to provide a sensible account consistent with the interpretations others made.

Finally, although no outcome is depicted in the example, the claims of the manager were interpreted by others at the hearing in terms of the context in which they were made, using the descriptive vocabularies as indexical expressions of sense-making practice. So the district manager's testimony can be interpreted as self-serving and evasive, giving the appearance that he was trying to escape responsibility for a decision.

Thus sense-making practices regulated discourse by requiring sensible and institutionally appropriate questions and responses within a context where counsel had the legal right to ask questions and expect responses. Sensible and institutionally acceptable responses necessitate the use of conventional and institutionally legitimate discursive resources and thereby further regulate or limit the kinds of statements that are made. In situations such as this one, where nonsensible and institutionally inappropriate responses emerge, one's behavior is interpreted as problematic and deficient. In general, sense-making practices allowed the board's views to prevail and supported the legitimacy of the board by establishing that the district manager breached the expectations that the legal counsel established using the sense-making practices. The manager held a unique and institutionally problematic perspective that led others to attribute responsibility to him for the unplanned fire, or at least for failure to avoid the fire.

As this abbreviated example suggests, ethnomethodological analysis is appropriate for examination of sense-making processes in actual social interaction related to regulatory discourse, and it also provides insights into sense making in documents and texts. As an analytic framework, ethnomethodology is especially appropriate for targeting the mutual negotiation and construction of meanings related to regulated discourse as key issues for examination, and it helps one to understand how particular interpretations were created and sustained. Ethnomethodological analysis of sense-making practices is less useful where the actions examined are not meaningful to subjects or where meanings are imposed on subjects and are nonnegotiable. Further, it is not as useful as narrative/rhetorical analysis in directly addressing the contents and forms of stories and narratives, although ethnomethodology can provide insights into the construction of narratives. In addition, ethnomethodology has been used only to a limited extent to address political and power dimensions of regulated discourse; it may thus be less useful in understanding the extent to which meanings advantage or disadvantage one group over another (Feldman, 1995). The next approach, Habermas's critical theory, is quite suitable for understanding the political and power related implications of hearing discourse and in showing how certain meanings advantage one group or person over another.

Habermas's Critical Theory

A third perspective, Habermasian critical theory (Alvesson & Wilmott, 1992; Gephart, Boje, & Thatchenkery, 1996; Gephart & Pitter, 1993; Habermas, 1973, 1979; Offe, 1984, 1985), provides insights into the ways that speech produces agreements that are considered legitimate by subjects. The theory necessarily connects speech acts in specific settings to broader political contexts and implications. That is, the theory addresses the nature of rational and valid speech acts and the ways that these speech acts support or undermine—legitimate or de-legitimate—macrosocial structures, including government agencies and other key social institutions. This perspective also examines how institutions regulate speech acts.

True legitimation in democracy requires policies based on social consensus developed through broad-based, democratic means (Gephart & Pitter, 1993; Kemp, 1985). For Habermas (1973, 1979), legitimation requires valid communication. Four types of speech acts form the ideal situation for valid communication, with each type corresponding to a distinct type of validity claim (Habermas, 1979; Kemp, 1985). *Communicatives* claim comprehensibility and require immanent statements and terms to be valid. *Representatives* claim truthfulness; they are validated by appeals to sincerity or truthfulness and are assessed in terms of assurances and consistency. *Regulatives* claim contextual appropriateness for utterances and are grounded in the convictions of participants. *Constative* claims are grounded in experiential sources and discourse that support a claim.

Habermas presumes everyday speech is based on a background consensus that emerges through ongoing recognition and reproduction of the four validity claims. According to Kemp (1985), where the claims cannot be immediately substantiated in discourse, participants must have recourse to some processes of mediation (p. 185). Free communication, Kemp explains, emerges only when all participants to a discourse have equal opportunity to use all four speech act categories (pp. 187-188). Thus the institutional conditions for consensus that serve general and not particular interests require a communication free from internal and external constraints. It is doubtful the ideal speech situation can be empirically realized because of political and psychological constraints on people. However, the ideal speech situation is a rational standard for judging existing discourses and can be used counterfactually as a measure of constraints on communication.

The political implications of speech acts are related to the production of institutional legitimacy because speech acts are the medium through which challenges to state and organizational legitimacy are created and refuted. Studies of legitimacy (Dowling, 1983; Gephart & Pitter, 1993) show that these challenges may be linked to a range of social issues and critical incidents. In these situations, government and other key social institutions use speech acts to create interpretations that will maintain their legitimacy. In public hearings, the need to maintain legitimacy is particularly acute because governmental priorities cannot be allowed to depend on the general formation of the public will that could contradict government and business needs. It is also difficult for government to ignore or address practical questions of stakeholders. So in public hearings, conflict is difficult to avoid or reduce.

Discourse analysis using the Habermasian framework is important to the integrated, critical/interpretive approach I am proposing because the framework contributes unique insights into how actual speech acts compare to an ideal-type speech situation in which, Pusey (1987) argues, "disagreements and conflicts are rationally resolved through a mode of communication which is completely free of compulsion and in which only the force of the better argument may prevail" (p. 73). A Habermasian framework provides an understanding of the political contexts and implications of speech acts, and it shows how discourse produces or challenges institutional legitimacy.

These features of critical theory and their relevance to understanding regulated discourse can be demonstrated by analysis of data concerning the pipeline fire inquiry discussed above (Gephart, 1992, 1993). In this inquiry, communicative speech acts that assert comprehensibility were regulated in several ways. For example, the hearing board set deadlines for interested parties to notify the board of their decision to participate, and the board advertised the inquiry in business sections of national and local newspapers. In addition, only parties with a legal interest were allowed to participate in the proceedings. Access to legal counsel, which facilitated inquiry participation, was readily available to corporate managers and government agencies that used legal counsel to present their views in a legally proper manner. For workers, legal counsel, was costly and hence problematic to secure. The judicial format restricted participants from questioning the board, counsel, or other participants, while the use of technical terms meant some statements were sensible only to professionals.

Representative speech acts that claim truthfulness were also regulated. The question-and-answer format limited testimony, and legal counsel determined when a question had been answered. Coached testimony that selectively presented facts was common among company witnesses. Significant constraints existed on what a party was willing to disclose, and external constraints influenced what they were compelled to disclose or withhold. Since simple yes or no answers to questions were common; the meaning of answers was limited by questions.

Regulative speech acts claiming contextual appropriateness grounded in convictions were also restricted. Differentiation of legal counsel and witnesses provided counsel with the authority to control conversations and to legitimately produce regulatives. Witnesses were compelled by oath to answer questions truthfully, but they could not ask questions. Further, the board had the authority to regulate all questions and testimony, to structure the inquiry, and to rule speech acts in or out of order. Proprietary information on the company was not made available to the public. Hence participants had unequal opportunity to provide statements at the inquiry.

Constative speech acts that claim truth, grounded in experiential sources, were also regulated. Truths were disclosed only as answers to questions. Factuality was assessed in terms of the proper use of technical language, and translation into technical language altered the character of events recounted. Witnesses were permitted to report only those facts relevant to the government energy board's mandate. Therefore, only certain truths were presented to the inquiry.

The inquiry thereby produced distorted communication that reflected the imbalances and inequalities of the inquiry itself (Kemp, 1985, p. 197). The consensus on facts that emerged at the inquiry was an apparent consensus, not a true one, since the prevailing interests were those of the state and capital. Communication, power, and organization emerged, in Mumby's (2001) terms, as interdependent and coconstructed phenomena (p. 585). Government and corporate stakeholders exerted power by regulating speech acts to privilege certain accounts and interpretations.

Habermasian critical theory is important to the integrated critical/interpretive approach I am proposing because it provides analytic concepts for assessing the political dimensions of communication. By emphasizing the role of language in mystifying power relations, the theory shows how the regulation of communication can produce decisions based on distorted understanding of events (Kemp, 1985). The theory thus provides insights that can help people live and work together more democratically. It reaffirms, as Kemp (1985) points out, the need for rational, truly democratic political decision making (p. 198).

As a part of an integrated approach for analyzing the discourse of public hearing and other institutions, Habermasian critical theory is especially valuable for analyzing the regulated dimensions of discourse and for assessing constraints on communication and processes that distort communication. The theory offers an effective means for examining how speech acts enhance or challenge personal and institutional legitimacy. The theory is also useful for understanding how one group dominates another using speech acts. As an analytic framework, Habermasian critical theory, however, is not well-suited for illuminating the forms and structures of narratives and stories. Also, because it relies upon ideal speech/act situations as the comparative basis for assessing actual situations, the theory is not especially useful for understanding regulatory sense making that departs from rational standards of interpretation.

THE CRITICAL/INTERPRETIVE APPROACH: INTEGRATING THE PERSPECTIVES

By combining the strengths of three separate perspectives, the critical/ interpretive approach provides a broad understanding of hearing discourse that examines narrative views of the world; the sense-making practices that construct meanings and world views in conversation; and the ways that narratives, sense-making practices, and speech acts produce grounds for rational and valid arguments that can reproduce or challenge institutional legitimacy. This integration of perspectives also facilitates insight about the various dimensions of regulation: the substance and form of stories of regulation, the practices and processes people employ to make sense of regulations and produce regulation through sense making, and the political aspects and implications of regulated communication.

The value of the integrated critical/interpretive approach can be demonstrated by examining further testimony of the district manager during the pipeline-accident hearing discussed earlier (Gephart, 1992, p. 127). This integrated analysis shows how accounts of local, situated logic were transformed into accounts consistent with top-down regulatory logic. Through this transformation process, the inquiry testimony was regulated and managed so as to remedy or at least avoid legitimation problems.

Counsel, Q1: "There would be less uncertainty if you flared than if you didn't flare? You don't agree with that?"
Manager, A1: "If I can use hindsight but with what I knew that day, no."
Counsel, Q2: "Well, in hindsight do you agree with that?"
Manager, A2: "In this particular case?"
Counsel, Q3: "In any case of an NGL leak, certainly a major leak?"
Manager, A3: "To go in and flare? I do not agree."
Counsel, Q4: "And that is on the basis that in your mind there is more certainty if you leave the gas unflared than if you flare it?"
Manager, A4: "As long as you monitor the cloud, yes."

Narrative/Rhetorical Analysis

This excerpt reveals that the meaning of the district manager's testimony is determined in large part by the narrative elements that precede this testimony. The testimony is part of the ongoing story of the fatal pipeline fire and the specific story of the district manager's problematic role in controlling the leak. The district manager narrates what happened to him and in so doing he constructs himself as a central character in the complicating actions and resolution of the story. Yet the district manager is initially placed in this role of central agent and villain by the questioning of legal counsel for the hearing board and by the actions and reactions of others present at the hearing.

The setting of the testimony provides further insight into its meaning to participants. The testimony of the district manager began during a special evening session of the hearing and continued the next day, and the direct and cross-examination of the manager comprised 352 pages of a 1,600-page transcript of the inquiry testimony. I attended the hearing and recall that there was an air of tension in the room during this testimony. Relatives of injured and deceased personnel displayed emotion; women relatives sobbed and dried their eyes using tissues pulled from boxes of tissue placed visibly near them. All eyes were on the district manager during his testimony, and the silence in the room was broken only by testimony and displays of emotion by the audience. Relatives of the injured and deceased were aware that the manager was not present at the leak site, and that he had admitted in earlier testimony that fires scared him. Thus observers questioned his motives and his competencies. His testimony was one of his last actions as district manager; he retired early from the company immediately after the hearing. These contextual factors allow us to see the discussion of flaring as an attempt, albeit unsuccessful, to save face and to protect his identity and role as district manager. Through testimony, the district manager is actively constructed as a poor manager. He was held responsible for the unplanned fire, and the process of constructing meaning occurred through narrativization, as well as through use of sense-making practices that are discussed next.

Ethnomethodological Analysis

This subsection addresses how sense-making practices are used to make top-down safety logics of regulatory agencies dominant in inquiry settings. Top-down safety logics implicitly support the legitimacy of key institutions and can be used to locate problems or deficiencies in other actors and agencies.

Past ethnomethodological research shows that two safety logics are operative in regulatory settings (Baccus, 1986). Top-down formal safety logic is produced in official accounts and specifies deductively logical conditions for preventing accidents of known kinds. It formulates overriding safety mechanisms including regulations sufficient to prevent accidents. Top-down safety logic reflects the mandated interests of regulators. It differs substantially from the actual situated safety logic that operators use and that relies on commonsense reasons for actions. Situated safety logic incorporates personal-safeguarding logic that allows emergent devices and ad hoc situational features to substitute for formal safety devices. Situated safety logic embraces the logic of object use whereby certain apparent demands emerge from the nature of the object. Informal practices based in situated safety logic are permitted by regulatory agencies if they work, but an accident is taken as direct evidence such informal practices have failed. Critical incidents such as accidents demonstrate a failure to provide safety and thereby threaten the legitimacy of the state, which is expected to regulate organizations and insure safety.

In my research (Gephart, 1992), I expected that top-down safety logic would characterize government regulatory agency discourse, and situated safety logic would be used by company personnel. I also expected the transformation of situated safety logic into top-down safety would be important for inquiry boards and inquiry legitimacy. The importance of these logics and their production by means of sense-making practices is evident through analysis of this testimony.

The counsel's questions construct a search for the cause or source of the decision not to flare the leak. The search targets prescribed steps that were missing during the efforts to control the leak and to troubles attached to the steps that were taken or missing (Gephart, 1992, p. 119, following Baccus, 1986). The counsel's skeptical questions (e.g., "You don't agree with that?") challenged the manager's explanation of the actions taken and the reasoning the manager employed. Regarding sense-making practices, the manager failed to create reciprocity in the perspectives that the board and he held. His conception of fires as safe and flaring as dangerous contradicts the formally accepted view of regulators that fires are dangerous, that flaring is the first step to create safety, and that stoppling is the second step. Counsel used an etcetera practice to provide the manager opportunities to revise his construction of the situated safety logic used to handle problems (Q1–Q4) and to emphasize the manager's faulty logic and interpretations. The indexical meaning of key terms was elaborated by the manager in a manner inconsistent with generally shared knowledge about natural gas liquids, leak control, and top-down regulatory safety logic. The testimony therefore shows that the district manager used a faulty approach with the wrong first step—don't flair, monitor—rather than the official approach of flaring then stoppling. The testimony therefore established that if the proper top-down safety logic and policies had been used rather than local safeguarding logic, the leak would have been flared and the accident avoided because it would not be deductively possible for an unplanned fire to occur if the leak were already ignited. The manager affirmed the utility of top-down logic, responding with the phrase, "If I can use hindsight . . ." to seemingly suggest the top-down logic was now validated. The ethnomethodological analysis thus shows how sense-making practices can be used to uncover the interpretations of the district manager concerning events and policies, and to show that these interpretations were based in a faulty situated logic rather than the more effective top-down logic of regulators that should have been used.

Critical/Theory Based Analysis

The sense-making practices used by counsel regulated communication and thereby undermined validity claims of the manager that appealed to situational logic (Gephart, 1992, pp. 130-131). By undermining the validity claims of the manager, the hearing board was able to enact procedural legitimation by publicly displaying that the government regulatory agency did have procedures and regulations sufficient to control organizations and prevent accidents (Gephart, 1992, p. 131). Undermining the manager's validity claims established that the manager deviated from formal policies and thereby preempted their effectiveness. It also demonstrated that this deviation was caused by the district manager and could not be readily anticipated or prevented by the board. For example, regarding communicatives, the manager failed to produce comprehensible responses to questions. The representative speech acts of the manager were undermined when counsel questioned the manager's sincerity and noted inconsistencies in his reasoning. Regulative speech acts that claimed the manager was a legitimate and effective supervisor were undermined by narratives, showing that the manager did not follow standard operating procedures for NGL leaks and that he did not answer inquiry questions correctly. Constative truth claims were undermined by the manager's failure to understand standard procedures for leak control. The public inquiry thus reproduced the legitimacy of the state and corporations by generating accounts that show how logics and validity claims used by the district manager diverged from mandated top-down logic to which pipeline company personnel were subject (Gephart, 1992, p. 132).

Integration

The integration of the three modes of analysis shows that the inquiry was designed and managed to remedy legitimation problems. The inquiry distorted accounts of local action and situated logic by converting these into accounts consistent with regulatory safety logics. Procedural legitimation was enacted by following formal inquiry procedures, by showing that the situated safety logic used by the manager was problematic, and by demonstrating that top-down regulatory logic, which specifies deductive conditions for preventing accidents, would have been effective if it had been used. Responsibility for the accident was allocated to the manager who failed to use the appropriate logic. The solution for prevention of future problems with similar leaks thus necessarily emerges from the logic of government regulatory discourse: one needs to ensure that company personnel use top-down safety logic to undertake sense making. The hearing board used sense-making practices to close off testimony once its own logic had been substantiated. Thus the accident and inquiry create the potentially incorrect impression that (1) key institutions are capable of regulating industry activities and ensuring safety and (2) the board functioned in a legitimate manner. The hearing board could do little to prevent the accident caused by defective sense making by a key manager and not by faulty regulatory action.

CONCLUSION

Regulation of communicative practices is a central feature of public hearings and inquiries. Governments need to manufacture the sense that social institutions are rational and fair. This perception of legitimacy is produced by systematically distorting inquiry communication so that dominant institutional logics prevail.

This chapter outlines a critical/interpretive approach to discourse analysis of public hearings and inquiries as regulated communication. This integrated approach can also be used to study other settings where regulated communication occurs, such as legal proceedings (Emerson, 1969; Garfinkel, 1967; Riddington, 1988, 1990) and meetings (Gephart, 1998; Schwartzman, 1993).

By integrating the analytic perspectives of narrative, ethnomethodology, and critical theory, the critical/interpretive approach underscores how communication, power, and meaning are interconnected by discourse, confirming Mumby's (2001) point that discourse constructs identities, experiences, and ways of knowing identities and experiences that serve certain interests over others (p. 595). The critical/interpretive perspective also encourages greater reflexivity in conceptualizing organizational power as regulation. It allows organizational studies to avoid reification of the organization by conceiving the organization as produced in and through communication.

Ultimately, the critical/interpretive approach to analyzing discourse offers insights into the centrality of communication to the cultural life of institutions. As this chapter has shown, narratives, sense-making practices, and speech acts are cultural practices that produce institutions and institutional regulation. As key settings for discourses that challenge and reproduce the legitimacy of social institutions, public hearings thus are an important example of cultural practices that regulate and are regulated by institutional and organizational activity.

ACKNOWLEDGMENTS

Research for this chapter is supported in part by funding from the Social Sciences and Humanities Research Council of Canada (SSHRC), whose support is gratefully acknowledged. I also wish to thank Mark Zachry and Charlotte Thralls for their suggestions and guidance in preparing this chapter. In addition, thanks to Claire McGuigan, Erich Welz, Tracy Yu, Paige Denham, and Michael Kulicki for research assistance.

REFERENCES

Alvesson, M., & Wilmott, H. (1992). *Critical management studies.* Thousand Oaks, CA: Sage.

Baccus, M. D. (1986). Multipiece truck wheel accidents and their regulation. In H. Garfinkel (Eds.), *Ethnomethodological studies of work* (pp. 20-59). Englewood Cliffs, NJ: Prentice-Hall.

Beck, U. (1992). *Risk society.* London: Sage.

Bogen, D., & Lynch, M. (1989). Taking account of the hostile native: Plausible deniability and the production of conventional history. *Social Problems, 36,* 197-224.

Boje, D. M. (2001). *Narrative methods for organizational and communication research.* Thousand Oaks, CA: Sage.

Boyer, W. W. (1960). Policy making by government agencies. *Midwest Journal of Political Science, 4,* 267-288.

Brody, H. (1981). *Maps and dreams.* New York: Pantheon Books.

Brown, A. D. (2000). Making sense of inquiry sensemaking. *Journal of Management Studies, 37,* 45-76.

Cicourel, A. V. (1973). *The social organization of juvenile justice.* New York: Wiley.

Cole, R. L., & Caputo, D. A. (1984). The public hearing as an effective citizen participation mechanism: A case study of the general revenue sharing program. *American Political Science Review, 78,* 404-416.

Coulon, A. (1995). *Ethnomethodology.* Thousand Oaks, CA: Sage.

Czarniawska, B. (1998). *A narrative approach to organization studies.* Thousand Oaks, CA: Sage.

Dowling, J. B. (1983). Legitimation, social structure and social order. In J. B. Dowling & V. MacDonald (Eds.), *The social realities of policing: Essays in legitimation theory* (pp. 1-51). Ottawa, Ontario: Minister of Supply and Services, Canada.

Emerson, R. (1969). *Judging delinquents: Context and process in juvenile court.* Chicago, IL: Aldine Publishing Company.

Emerson, R. (1981). On last resorts. *American Journal of Sociology, 87,* 1-22.

Fairclough, N. (1992). *Discourse and social change.* Cambridge, UK: Polity Press.

Feldman, M. (1995). *Strategies for interpreting qualitative data.* Thousand Oaks, CA: Sage.

Garfinkel, H. (1967). *Studies in ethnomethodology*. Englewood Cliffs, NJ: Prentice-Hall.

Gephart, R. P. (1992). Sensemaking, communicative distortion, and the logic of public inquiry legitimation. *Industrial Crisis Quarterly, 6,* 115-135.

Gephart, R. P. (1993). The textual approach: Risk and blame in disaster sensemaking. *Academy of Management Journal, 36,* 1465-1514.

Gephart, R. P. (1997). Hazardous measures: An interpretive textual analysis of quantitative sensemaking during crises. *Journal of Organizational Behavior, 18*(special issue), 583-622.

Gephart, R. P. (1998). Status degradation and organizational succession: An ethnomethodological approach. In J. Van Maanen (Ed.), *Qualitative studies of organization* (pp. 159-191). Newbury Park, CA: Sage.

Gephart, R. P., & Pitter, R. (1993). The organizational basis of industrial accidents in Canada. *Journal of Management Inquiry, 3,* 238-252.

Gephart, R. P. (2004). Sensemaking and new media at work. *American Behavioral Scientist, 48*(4), 479-495.

Gephart, R. P., Boje, D. M., & Thatchenkery, T. (1996). Postmodern management and the coming crises of organizational analysis. In D. M. Boje, R. P. Gephart, & T. Thatchenkery (Eds.), *Postmodern management and organization theory* (pp. 1-18). Thousand Oaks, CA: Sage.

Habermas, J. (1973). *Legitimation crisis*. Boston, MA: Beacon Press.

Habermas, J. (1979). *Communication and the evolution of society*. Boston, MA: Beacon Press.

Habermas, J. (1989). *The transformation of the public sphere: An inquiry into a category of bourgeois society*. Cambridge, MA: MIT Press.

Heap, J. (1975). What are sensemaking practices? *Sociological Inquiry, 46,* 107-115.

Keeling, A. (2001). The rancher and the regulators: Public challenges to sour-gas regulation in Alberta, 1970-1994. In R. Epp & D. Whitson (Eds.), *Writing off the rural west: Globalization, governments and the transformation of rural communities*(pp. 279-300). Edmonton, AB: University of Alberta Press and the Parkland Institute.

Kemp, R. (1985). Planning, public hearings and the politics of discourse. In J. Forrester (Ed.), *Critical theory and public life* (pp. 177-201). Cambridge, MA: MIT Press.

Leiter, K. (1980). *A primer in ethnomethodology*. New York: Oxford University Press.

Lynch, M. (1993). *Scientific practice and ordinary action: Ethnomethodology and social studies of science*. Cambridge, UK: Cambridge University Press.

Lynch, M., & Bogen, D. (1996). *The spectacle of history: Speech, text and memory at the Iran-Contra Hearings*. Durham, NC: Duke University Press.

Mumby, D. (2001). Power and politics. In F. Jablin & L. Putnam (Eds.), *The new handbook of organizational communication* (pp. 585-623). Thousand Oaks, CA: Sage.

Nikiforuk, A. (2001). *Saboteurs: Wiebo Ludwig's war against oil*. Toronto: McClelland & Stewart, Ltd.

Offe, C. (1984). *Contradictions of the welfare state*. Cambridge, MA: MIT Press.

Offe, C. (1985). *Disorganized capitalism*. Cambridge, UK: Polity Press.

Pusey, M. (1987). *Jürgen Habermas*. Chichester, UK: Ellis Harwood, Ltd.

Putnum, L., & Fairhurst, G. (2001). Discourse analysis in organizations: Issues and concerns. In F. Jablin & L. Putnam (Eds.), *The new handbook of organizational communication* (pp. 78-136). Thousand Oaks, CA: Sage.

Riessman, C. K. (1993). *Narrative analysis*. Thousand Oaks, CA: Sage.

Riddington, R. (1982). When poison gas comes down like a fog: A native community's response to cultural disaster. *Human Organization, 41,* 36-42.

Riddington, R. (1988). *The trail to heaven: Knowledge and narrative in a native northern community*. Vancouver, BC: Douglas & McIntyre, Ltd.

Riddington, R. (1990). *Little bit know something: Stories in a language of anthropology*. Vancouver, BC: Douglas & McIntyre, Ltd.

Salter, L., & Slaco, D. (1981). *Public inquiries in Canada*. Ottawa: Science Council of Canada.

Schwartzman, H. B. (1993). *Ethnography in organizations*. Thousand Oaks, CA: Sage.

Trice, H. M., & Beyer, J. M. (1993). *The culture of work organizations*. Englewood Cliffs, NJ: Prentice-Hall.

Turner, R. (1974). *Ethnomethodology*. Harmondsworth, UK: Penguin.

Turner, B. A. (1976). The organizational and interorganizational development of disasters. *Administrative Science Quarterly, 21,* 378-397.

Turner, B. A. (1978). *Man-made disasters*. New York: Crane, Russack.

Vaughan, D. (1996). *The Challenger launch decision: Risky technology, culture and deviance at NASA*. Chicago, IL: University of Chicago Press.

Wynne, B. (1980). Technology, risk, and participation: On the social treatment of uncertainty. In J. Conrad (Ed.), *Society, technology and risk* (pp. 83-107). London: Academic Press.

Contributors

David M. Boje is professor of Management and the Arthur Owens Chair in Business Administration at New Mexico State University. In addition to extensive journal publications, he is the author or editor of several books, including *Narrative Research Methods for Communication Studies, Postmodern Management and Organizational Theory*, and *Managing in the Postmodern World: America's Revolution Against Exploitation*. His research interests include globalization, empowerment, and organizational behavior.

David Clark is assistant professor in the Department of English at the University of Wisconsin, Milwaukee, where he teaches in the graduate and undergraduate programs in professional and technical writing. His research focuses on the rhetoric of technology, knowledge management, multimedia documentation, and interface design. His current publications include articles in *Technical Communication Quarterly* and *Business Communication Quarterly* and a chapter in *Computers and Technical Communication*.

Brenton Faber is associate professor of Communication and Media at Clarkson University. He investigates the operation of discourse in contexts of social change. His research has appeared in a variety of rhetoric and business and technical communication journals. He is also the author of *Community Action & Organizational Change: Image, Narrative, Identity*. Currently he is working on funded research in the public representations of nanoscale science and technology.

Robert P. Gephart, Jr., is professor of Strategic Management and Organization in the School of Business at the University of Alberta. His research has appeared in the *Academy of Management Journal, Administrative Science Quarterly, Journal of Management, Journal of Management Inquiry, Journal of Organizational Behavior*, and *Qualitative Sociology and Sociological Perspectives*. He is the author of *Ethnostatistics: Qualitative Foundations for Quantitative Research* and a co-editor of *Postmodern Management and Organization Theory* (with D. Boje and T. Thatchenkery). His current research addresses sense-making related to technological and environmental risks.

Kenneth J. Gergen is Mustin Professor of Psychology at Swarthmore College and affiliate professor at Tilburg University in The Netherlands. He is a leading international scholar in social construction theory. As author or editor of more than 25 books and numerous articles, Gergen has written extensively on social construction, exploring its connections to such wide-ranging topics as pedagogy, identity and community, technology, relational theory and the self, aging, education, and organizations. Recent publications include *Relational Responsibility* (with S. McNamee), *An Invitation to Social Construction, Social Construction in Context, The Appreciative Organization* (with H. Anderson, D. Cooperrider, S. McNamee, and D. Whitney), and *Social Construction, a Reader.*

Carl G. Herndl is associate professor of Rhetoric and Professional Communication at Iowa State University. His work has been widely published in rhetoric and composition journals, and he served as guest editor for a special issue of the *Journal of Business and Technical Communication* on critical practice in professional communication. He was also co-editor of *Green Culture: Environmental Rhetoric in Contemporary America,* which received the 1997 National Council of Teachers of English award for the Best Collection in Scientific and Technical Communication. His areas of research include cultural studies, rhetorical studies of science, and theories of rhetorical and cultural agency.

Adela C. Licona is assistant professor of Rhetoric and Professional Communication at Iowa State University. Her research interests include women's studies, chicana theory, cultural studies, and feminist rhetorics. She is co-editor of the forthcoming collection *Reading, Writing, Women: Historical Practices and Current Conversations* (Indiana University Press).

Lorelei Lingard is associate professor in the departments of Paediatrics and Health Policy, Management, and Evaluation at the University of Toronto, scientist at the Hospital for Sick Children Research Institute, and educational scientist at the Centre for Research in Education. She is co-editor of *The Rhetoric and Ideology of Genre: Strategies for Stability and Change.* Her research interests include inter-professional team communications in health care, education of novice health care professionals, relationships between team communications and patient safety, linguistic pragmatics, and rhetorical analysis.

Wanda Orlikowski is professor of Information Technologies and Organization Studies at the Massachusetts Institute of Technology (MIT) Sloan School of Management and the Eaton-Peabody Chair of Communication Sciences at MIT. She is the senior editor of *Organization Science.* Her research examines organizational changes associated with the use of information technology and includes extensive published studies of groupware technologies and electronic media in organizations. Orlikowski is currently head of a five-year project (funded by the National Science Foundation) on the social and economic implications of Internet technology use in organizations.

Martin Ruef is associate professor of Sociology at Princeton. He is the co-author of *Institutional Change and Healthcare Organizations*, which won the Max Weber prize from the American Sociological Association. His research interests include organizational theory, economic sociology, network analysis, and the sociology of culture.

Barbara Schneider is associate professor of Communication and Culture at the University of Calgary. Her research areas include the social and cultural aspects of schizophrenia and communication in organizational settings. She is the author of numerous research articles in rhetoric, professional communication, and health journals, and is currently conducting a funded participatory research study of housing for people with severe mental illnesses.

Catherine F. Schryer is associate professor in the Department of English Language and Literature at the University of Waterloo. Widely published in such journals as *Journal of Business and Technical Communication, Written Communication, Medical Education*, and *Social Science and Medicine*, her research focuses on genre theory with a special emphasis on health communication genres. In addition to winning numerous research grants, she is the recipient of the 2001 National Council of Teachers of English award for Best Article Reporting Qualitative or Quantitative Research in Scientific and Technical Communication.

Marlee Spafford is associate professor at the University of Waterloo's School of Optometry. She holds a Doctor of Optometry degree and M.Sc. (Physiological Optics) from the University of Waterloo and a Ph.D. (Theory & Policy Studies in Education) from the Ontario Institute for Studies in Education/University of Toronto. She is a Fellow of the American Academy of Optometry. Her research interests include health care professional education, communication, socialization, and professional gate-keeping.

Clay Spinuzzi is associate professor of Rhetoric in the Division of Rhetoric and Writing at the University of Texas at Austin, where he directs the Computer Writing and Research Lab. His research focuses primarily on sociocultural approaches to investigating, evaluating, and improving information design; it has been published in a wide range of journals and anthologies. Spinuzzi was the 2003 recipient of the National Council of Teachers of English (NCTE) award for Best Article on Philosophy or Theory in Scientific and Technical Communication, and in 2004 his book *Tracing Genres Through Organizations: A Sociocultural Approach to Information Design* received the annual prize for Best Book in Scientific or Technical Communication by NCTE.

Dorothy Winsor is professor of Rhetoric and Professional Communication in the Department of English at Iowa State University. She is an award-winning author for her ethnographic studies on the writing of engineers in industry and has won four National Council of Teachers of English awards in scientific and technical communication; she was also recipient of the 2003 Association for Business Communication Award for Distinguished Publication in Business Communication. Her two books are *Writing Like an Engineer: A Rhetorical*

Education and *Writing Power: An Ethnographic Study of Writing in an Engineering Center*. Winsor is also editor of the *Journal of Business and Technical Communication*.

JoAnne Yates is Sloan Distinguished Professor of Management at the Massachusetts Institute of Technology (MIT) Sloan School of Management. Her interdisciplinary program of research includes historical and contemporary studies of communication and information technology in American firms. She is the author of *Control Through Communication: The Rise of System in American Management* (Johns Hopkins University Press, 1989) and *Structuring the Information Age: Life Insurance and Technology in the Twentieth Century* (Johns Hopkins University Press, 2005). Yates is the recipient of numerous publication awards, including the Harold F. Williamson, Sr., Medal for mid-career excellence in Business History and the Association for Business Communication Award for Distinguished Publication in Business Communication.

Index

Journal of the American Medical Association, 103
Jurisdictional claims. *See* Argumentation among health-care professionals

Kairos, 32, 134, 135
  See also Agency and the possibilities of social action
Kenobe, Inc., 232
Kensington Consultation Centre, 225
Kinder, Richard, 230
Knowledge-generating cultures, processes characterizing, 113-114
Kuhn, Thomas, 114

Labor contracts. *See* Texts used to manage continuity/change in an activity system
Laclau, Ernesto, 136-137
Language and Power (Bourdieu), 25
Latour, Bruno, 50
Lay, Kenneth, 230, 234
Legal force and texts used to manage continuity/change in an activity system, 6
Legitimacy/public support and critical/ interpretive approach to public hearings/inquiries, 243-245, 253-256, 259-260
Lesbian groups and theater as a vehicle of scholarly expression, 126
"Level of Discourse Continues to Slide, The" (Schwartz), 76
Liberation Management: Necessary Disorganization for the Nanosecond Nineties (Peters), 158
Linguistic analysis and critical text analysis in workplace research, 205-207
Linux, 159
Little, Brown Handbook, The (Clifford), 136
LJM partnerships, 232
Logical coherence and writing/ relationship in academic culture, 116-117
Logical conformism, 26

Logic of Practice, The (Bourdieu), 138
Longo, Bernadette, 136, 160
Lunsford, Andrea, 139

Mailloux, Steven, 143
Mark, Edward, 88
Marvin, Carolyn, 157
Marxist theory, 136
Mediation and self-regulation, 46-47
Medical Care (Lewis), 97
MEDLINE, 102-103
Memo of Understanding as a subsidiary document, 15, 16
Memos and genre theory, 71
Messer-Davidow, Ellen, 135, 143-145
Microsoft
  Office Suite, 75
Microsoft Corporation, 68
Midsummer Night's Dream, 234
Miettinen, Reijo, 51
Milbank Memorial Fund Quarterly, 103
Miller, G., 189
Miller, L., 187
Missing in Action, 157
Modern Healthcare, 103
Morris, Meagen, 139
Mouffe, Chantal, 136-137
Mulkay, Michael, 120-121

Narrative/rhetorical analysis and critical/ interpretive approach to public hearings/inquiries, 240, 244, 245-249, 257
Narrative studies, the antenarrative turn in
  antenarrative theory, 223-225, 234-235
  deconstructionists, 219-220
  dialectic theory, 225-226
  Enron, 226-236
  exclusion of story from narrative analysis, 220-223
  Gas bank project at Enron, 230-234
  implications of antenarrative theory, 234-235
  overview, 219-220
  realism/structuralism/social constructionism, 219